AGRARIAN SYSTEM OF MEDIEVAL ASSAM

About the Author

Dr. (Mrs.) Jahnabi Gogoi (Nath) (*b*. 1960) has brilliant academic career all throughout. An M.A. in medieval history from Dibrugarh University, she served for sometime as lecturer in Chabua College, Chabua, till she joined the History Department of her parent University in 1987. She was awarded Ph.D. Degree by Dibrugarh University in 1994. Dr. Gogoi has to her credit a number of highly acclaimed research papers on economic history and agrarian life of pre-colonial Assam.

Dr. (Mrs.) Gogoi (Nath) is a member of the All-India History Congress, the North-East India History Association and the Dibrugarh University Women's Forum.

AGRARIAN SYSTEM OF MEDIEVAL ASSAM

Dr. (Mrs.) JAHNABI GOGOI (NATH)
Department of History, Dibrugarh University
Dibrugarh (Assam)

CONCEPT PUBLISHING COMPANY, NEW DELHI-110059

ISBN 81-7022-967-7

First Published 2002

Published and Printed by
Ashok Kumar Mittal
Concept Publishing Company
A/15-16, Commercial Block, Mohan Garden
NEW DELHI-110059 (India)

Phones : 5351460, 5351794
Fax : 091-11-5357103
E-mail: publishing@conceptpub.com

PREFACE

Despite the formation of the ancient state of Pragjyotisha-Kamarupa in the lower Brahmaputra valley and its existence till at least the middle of the 12th century A.D., political integration of the entire valley was not achieved till the coming of the Ahoms and their establishment of a state here subsequently known as Assam. The Ahoms who came to this valley from the region of Upper Burma and southern Yunan in the beginning of the 13th century formed the nucleus of a small state in the extreme south-eastern part of the Brahmaputra valley and by way of its gradual expansion by subjugating the local tribes, made it a considerably big state so that by the late 17th century it came to cover almost the entire region of the Brahmaputra valley except the extreme western part now represented by the undivided district of Goalpara.

The Ahoms ruled the land till the end of the first quarter of the 19th century covering a period of about six hundred years. During this period the ruling class by way of their inherited proficiencies and follow-up policies of administration systematically utilised the available man-power and converted the jungle lands of Assam into open and extensive cultivable fields; and thereby helped in shaping the socio-economic structure of the land into a consolidated whole which till date characterises the society of Assam. Predominently characterised by the tribal way of living, the economy of the upper Brahmaputra valley was almost revolutionised with the introduction and extension of the permanent wet-rice cultivation. Not being accustomed to the caste systems at the initial stage, which was the characteristic mostly of the central society of the lower Brahmaputra valley in the period, the Ahoms soon realised the role of caste hierarchy in socio-economic and cultural life and adopted policies to honour it in Assam's agrarian society. This was why both caste and tribe characters remained in the agrarian society of Assam in medieval period.

The Ahom rulers were adopting a policy of isolation to which Assam's geographical location also largely contributed. This needed the growth of a self-sufficient economy in Assam in the medieval period. As a result, peasants in Assam in the period under review had little knowledge of the out-side world and whatever they produced they remained content with them. But after their contact with the Mughals this isolation slowly melted, although modernisation of the agrarian society of Assam came only after the British occupation of the land.

Methods and appliances of production introduced by the ruling Ahoms were comparatively advanced to those of the local tribes. But the system of their manpower utilisation was feudalistic. Industry, trade and money circulation were limited. Distinctions between man and man on the basis of caste and political status were maintained in socio-cultural life, which led to the growth of a privileged class in the society. Thus, slowly social classes based on the ritual status of caste, and political and economic standard appeared in the agrarian social structure, which, however, led to socio-economic transition in the state towards the close of our period.

A few works dealing with certain aspects of Assam's agrarian society of medieval period have been done; but no serious and systematic study covering specially the agrarian structure has yet been undertaken by any scholar. J.N. Phukan's *Economic History of Assam Under the Ahoms*, is undoubtedly a pioneering work in the field; but it covers a wide spectrum of the economy of Assam of medieval times. The present study attempts to concentrate on the agrarian structure of Assam of the period only. We have tried here to make a scientific approach to the investigation of the subject on the basis of all possible and available source materials. We, however, admit our shortcomings while interpreting things and drawing conclusions.

Major sources of our information are the local and contemporary chronicles (*buranjis*), land grant epigraphs, biographical and genealogical works of medieval period of Assam (both published an unpublished), the contemporary foreign accounts and the Persian chronicles. Besides, we have used a number of British records which were written either in the last part of the Ahom rule in Assam or immediately after the fall of the kingdom. It is to be noted that most of such early British records were prepared after proper field investigations. Such records were supplemented by a few of the petitions submitted by the local gentry to their British administrators. Secondary sources include published and unpublished books, articles and journals. It is a fact that a major part of the sources used here is more appropriate to the study of the late medieval systems than the early medieval ones; but every possible attempt has been made at the same time, to collect and use materials pertaining to the study of the earlier systems also.

While collecting our materials we had to approach many of the educational and research institutions, like Universities of Dibrugarh,

NEHU and Gauhati, the Department of Historical and Antiquarian Studies (Guwahati), the Assam Secretariat Record Room (Guwahati), the National Library (Calcutta), the Library of the Indian Council for Social Sciences Research, Shillong, and a number of expert individuals. We acknowledge our gratefulness to all those institutions and individuals.

I take this opportunity to acknowledge my gratefulness to Dr. (Mrs.) S.L. Barua, Professor of History, Shri Shri Aniruddhadeva Chair Dibrugarh University, who kindly supervised my work when it was prepared as a thesis for Ph.D. Degree of Dibrugarh University. I also offer my sincere regards to my husband Dr. D. Nath, Professor & Head, Department of History, Dibrugarh University, for all co-operations he had accorded to me in completing the work. Thanks are also due to Prof. J.B. Bhattacharjee, Ex-Vice-Chancellor of the Assam University, Silchar, and Prof. J.N. Phukarn, Retired Professor and Head of the History Department, Guwahati University, for their comments and suggestions as examineer of my work, which are taken into consideration before it was placed for publication.

Dibrugarh University
Dibrugarh.
August 2001

Mrs. JAHNABI GOGOI (NATH)
Reader,
Department of History

CONTENTS

ABBREVIATIONS

AAH	*An Account of Assam* by Francis Hamilton.
AB	*Ahom Buranji*, (tr. & ed.) G.C. Barua.
ABHB	*Asam Buranji* by Harakanta Sarma Barua Sadaramin.
ABGB	*Asam Buranji* by Gunabhiram Barua.
ABS	*Asam Buranji Sar*, (ed.) P.C. Choudhury.
ABSM	*Asam Buranji* obtained from the family of Sukumar Mahanta, (ed.) S.K. Bhuyan.
ADB	*Ahomar Din* by Hiteswar Barbarua.
Baharistan	*Baharistan-i-Ghaybi* (tr. & ed.) M.I. Bora.
CHAB	*A Comprehensive History of Assam* by S.L. Barua.
DAB	*Deodhai Asam Buranji*, (ed.) S.K. Bhuyan.
DAR	*A Descriptive Account of Assam* by W. Robinson.
DHAS	Department of Historical and Antiquarian Studies, Guwahati.
DRV	*Darrang Raj Vamsavali*, (ed.) N.C. Sarma.
MSMCD	*Mahapurusha Sri Sankaradeva Aru Sri Madhavadeva Charita* by Daityari Thakur.
HAG	*A History of Assam* by E.A. Gait.
Habib (B)	*Mughal Bharater Krishi Vyavastha* being the Bengali translation of *The Agrarian Systems of Mughal India 1556-1707* by I. Habib, (tr.) Dutta, Pain & Palit.
JASB	*Journal of Asiatic Society of Bengal.*
JARS	*Journal of Assam Research Society.*
JHR	*Journal of Historical Research*, Dibrugarh University.
KGC	*Katha Guru Charit*, (ed.) U.C. Lekharu.
MEAG	*Medieval and Early Colonial Assam* by A. Guha.
PAB	*Purani Asam Buranji*, (ed.) H.C. Goswami.
PSN	*Prachya Sasanawali* by M. Neog.
Riyaz	*Riyaz-us-Salatin*, (tr.) Abdus Salam.
SAB	*Satsari Asam Buranji*, (ed.) S.K. Bhuyan.
TB	*Tungkhungia Buranji*, (ed.) S.K. Bhuyan.
TB (Eng.)	English version of the *Tungkhungia Buranji* (ed. & tr.) S.K. Bhuyan.

GLOSSARY

Agar/Agaru	— a kind of tree (*acquilaria agallocha*) from which a kind of perfume is prepared.
Ahu	— a kind of paddy sown broad-cast in winter and harvested in summer.
Ali	— road/embankment.
Bahatiyas	— peasant-*paiks* residing in the foot-hill areas of Assam who were allotted to the hill tribes by the Assam rulers for rendering service or to pay a stipulated tribute in kind called *posa* to their hill masters.
Bar-kapor	— a coarse variety of woven cotton cloth used as a wrapper.
Barangani	— subscription; a tax levied on the *Satras*.
Bhakat	— a Vaishnava bachelor devotee of a *Satra*.
Bigha	— unit of land measurement equal to 14,400 sq.ft.
Bihu	— national festival of the Assamese. There are three Bihus in Assam: the Bahag Bihu or Rangali Bihu, the Kati Bihu or Kangali Bihu, the Magh Bihu or Bhogali Bihu.
Bil	— a pond, a small lake.
Brahmottar	— rent-free lands granted to the Brahmanas.
Buranji	— chronicles written in Assamese or Tai.
Cowri	— a shell used as coin in early and medieval Assam.
Chadar or Cheleng	— a piece of cloth used as upper garment by the Assamese.
Changmai	— Ahom royal cook.
Charit/Charit Puthi	— biography of the medieval Vaishnava saints of Assam.
Choky	— an out-post, a customs house.
Churiya	— a waist cloth of a man; also called *dhuti*
Dao	— a long knife or a short sword with a long handle.
Dargah	— a sacred burial place of a *pir*.
Dasa	— slave

Devottar	— revenue-free lands gifted for the erection and maintenance of temples.
Dewal	— the second man of a *got* of *paiks*.
Dharmottar	— revenue-free lands gifted for religious purposes.
Dhuti	— loin cloth of a man, a garment for the lower part of the body of a man.
Dola	— Assamese litter, sadan or palanquin.
Don	— unit of measure of weight of rice, paddy, pulses etc.
Duar	— a mountain pass.
Faqir	— a mendicant or ascetic.
Ga-dhan	— poll tax.
Gandhiya Bharal	— library and store for official records and chronicles of the Ahom rulers.
Ghat	— landing place on the bank of a river, a tax for using such a place for trade.
Gomasta	— a leader or a chief of a department under the Koches.
Got	— a unit of four or three peasant *paiks*.
Guru	— a preceptor.
Hat	— market place, a periodical market.
Hatkhowa	— a lessee or collector of taxes in a *hat* or market.
Hatimur	— a unit of one thousand *paiks*.
Jangphai	— amber.
Japi	— a broad, circular bamboo-made cover used as sun-shade or in rain.
Karkhana	— workshop of a Bhuyan; the court or the assembly under a Bhuyan.
Khat	— an estate of the nobles in medieval Assam.
Likchou	— a *paik* assigned to the nobles for their private use.
Mon	— unit of measurement of weight, usually of 40 *sers* or 37 kgs.
Moamariya	— disciples of the Moamariya or Mayamara *satra*.
Mul	— the first man of a *got* of *paiks*.
Nankar	— rent-free lands granted to persons other than the Brahmanas.
Pamua	— temporary cultivators.
Pandit	— a scholar, a teacher.
Pirpal	— rent-free lands granted to the Muslim saints called *Pir* or to a mosque.

Piyal — census.
Pura — unit of measure of lands, rice, pulses etc.
Riha — upper garment of woman.
Satra — Vaishnava monastery.
Satradhikara — head of a Vaishnava monastery.
Ser — a unit of weight about 930 gms. or 1/40 of a *mon*.
Tewal — the third man of a *got* of *paiks*.
Tola — a measure of weight equal to 1/80th part of a *ser*.
Vamsavali — a genealogical work of the Vaishnava saints or the Koch kings.
Wazir — an officer, minister in the Muslim Administration.

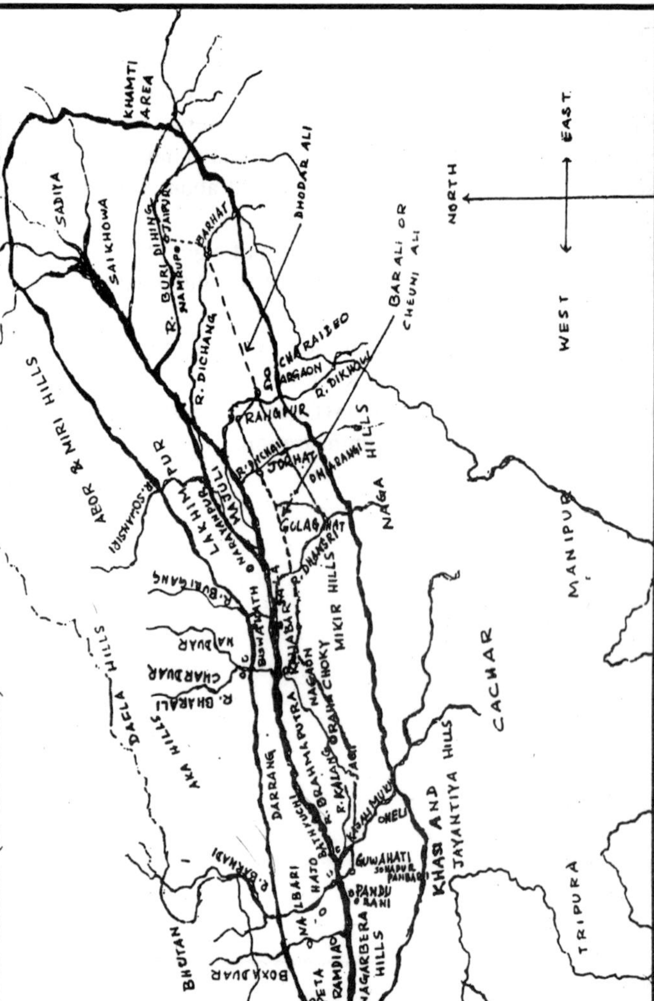

Assam in Medieval Times (*in its zenith*)

1

INTRODUCTION

Scope and Objective of the Work

'Asam'—the name by which the valley of the Brahmaputra came to be known during the rule of the Ahoms,[1] covered in medieval times, roughly the region from Sadiya in the east to the river Manah in present Kamrup district in the west on both banks of the Brahmaputra, excluding of course, the surrounding hill tracts. After the fall of the ancient kingdom of Pragjyotisha-Kamarupa, in the middle of the 13th century, which covered the western part of the kingdom of Asam, the region witnessed the growth of a number of states formed by the local tribes. These included the kingdom of the Chutiyas and that of the Kacharis, besides smaller areas dominated by the Maran and the Barahi tribes,[2] and the Bhuyans.[3] In the beginning of the second quarter of the 13th century, a small kingdom was established by the Ahoms, a branch of the Tais, who came from upper Burma, in the extreme south-east, while another tribe—the Koches established their kingdom in the extreme west by the beginning of the 16th century. Like the kingdom of Magadha which rose to power by exterminating other contemporary kingdoms and principalities[4] in the Gangatic valley, in the Brahmaputra valley also, the Ahom kingdom finally emerged as the only single state and covered the entire areas by extinguishing the other tribal states and the Bhuyan territories. Indeed by middle of the 17th century the Ahom kingdom, then known as *Asam Desh* or *Asam Rajya*,[5] became the single largest state establishing its suzerainty over the entire area. This the rulers of Asam had done not only through their military might and political sagacity but also by an advanced mode of production accompanied by a new type of land relations.

The scope of the work is confined to the study of agrarian history of Assam of the period from the first half of the 13th century to the first quarter of the 19th, when Asam was taken over by the British after the treaty of Yandaboo in 1826 A.D. It was, therefore, a period of about six hundred years when the Ahoms were ruling over the state. The territorial unit which was then known by the name 'Asam', covered almost the entire Brahmaputra valley excluding Goalpara, which was never a part of the state of 'Asam' in medieval times; and, therefore, our study will be confined to the territories now covered by Assam proper excluding Goalpara and the Barak valley. The period is important in the history of Assam in the sense that it was during this period that, the state of 'Asam' was formed and its society was shaped both of which came to be identified subsequently as the state of Assam and the Assamese society respectively.

The objective of our study is to find out the forces and factors that were at work to form the Assamese agrarian society during our period. The study of the sources of economy, its nature, production agents, their relations, volume of production and its distribution, development of trade and its extent are taken as some of the major aspects behind the formation of an agrarian society. We shall therefore concentrate our study over those aspects which are characteristics of such a society. It is to be mentioned that since very ancient times, although Assam has been maintaining an inherent link with India in the aspects of its cultural developments, yet in most of the cases, it still retains its distinct identity. In our period of discussion the agrarian society of Assam which was being formed, possessed certain specifications, which made its agrarian life distinct from that of contemporary India. But no systematic study in this field has yet been done by any scholar. Stray references to its economic conditions have been found in the works of certain scholars like U.N. Gohain's *Assam under the Ahoms* (1942), N.K. Basu's *Assam in the Ahom Age* (1970) and J. Bargohain's *Asamar Arthanaitik Itihas* (Assamese, 1985). Some more valuable works in this line have been done by Amalendu Guha in his various Assamese and English articles published in different research journals during the period from the late sixties to the eighties. Of his selected articles, a collection with the title *Medieval and Early Colonial Assam* (1991), has been published. S.L. Barua in a number of her articles published in different research journals and also in her *A Comprehensive History of Assam* (1985), has dealt with some of the agrarian aspects of medieval Assam. One of the pioneering works on the economy of Assam of the medieval period is that of J.N. Phukan, entitled *The Economic History of Assam under the Ahoms* (1973), a Ph.D. thesis of Gauhati University covering wide spectrum of economic history. In recent years S.N. Sarma's *Socio-*

Economic and Cultural History of Medieval Assam (1989) and S. Rajguru's *Medieval Assamese Society* (1988), have been published; but both of the works are not dealing with the agrarian life of Assam of the period in proper sense of the term. Outside Assam, comprehensive works on this subject on the period of the Muslim rule in India has been done by scholars like Moreland in his *The Agrarian System of Moslem India* (1929), Irfan Habib in *The Agrarian System of Mughal India (1556-1707)* (1963), and also on neighbouring Bengal by A.C. Banerjee in *Agrarian System of Bengal* (2 Vols. 1980). But no such works have yet been done on Assam. Under such circumstances, the present project has been undertaken to make a detailed and systematic study of Assam's agrarian system of medieval times.

Best efforts have been made to make the method of treatment on the subject scientific and analytical and taking into account all the available sources. There are stores of materials in the forms of both literary and epigraphic records including numismatic. Besides, accounts of the contemporary foreign travellers throw light on the subject. All these sources are carefully scrutinised and their authenticity examined. Conjectures have been made when necessary and conclusions drawn thereon with the help of the oral literature and by studying certain customs and traditions, which still exist in some villages of Assam, for cases where there are no direct evidences in contemporary records.

Historical Background

In order to understand the agrarian system of medieval Assam and that of the tribes under the jurisdiction of the Ahom hegemony, a study of the emergence of the Ahom kingdom as the single largest state in the Brahmaputra valley is necessary. After the fall of the Palas, the ancient kingdom of Pragjyotisha-Kamarupa declined towards the middle of the 12th century A.D. The process of its collapse was almost complete when its king, possibly one Sandhya (c. 1250-1260 A.D.), shifted his capital city from Kamarupanagara (modern north Guwahati) to Kamatapur in Koch Behar about the year 1260 A.D. Since then, a new kingdom known after its capital city, as the kingdom of Kamata or Kamrup-Kamata came into being, which lasted from the middle of the 13th century to the close of the 15th century over the region now covered by the districts of Kamrup and Goalpara (undivided) in Assam, and Koch Behar in North Bengal. During this period of its existence, a number of rulers belonging to different dynasties ruled the kingdom.[6] In the reign of its last ruler Nilambar, who ruled till 1498 A.D. Ala-ud-din Hussain Shah of Bengal (1493-1519 A.D.) invaded and occupied the kingdom, established within it a Muslim colony and appointed his son Danial as its ruler.[7] The Persian sources mention that Hussain Shah subjugated also a number of Bhuyans or local land-lords like Rupnarayan, Gosal Khan, Malkonwar and Lakhinarayan[8] and established his hold over the entire region as far east as the river Barnadi.

The fall of the Kamata kingdom synchronised with the peculiar political situation throughout the entire Brahmaputra valley. With the fall of the Kamata kingdom, even the remnants of the old kingdom of Pragjyotisha-Kamarupa extinguished. A process of disintegration set in, paving the way for the growth of a number of small principalities formed mainly by tribes. While in the region covered by the Kamata kingdom, there arose the kingdom of the Koches, a tribe of Mongoloid origin, in the eastern part of the valley a number of other Mongoloid tribes formed their states, the most notable among them being the Chutiyas, the Kacharies and finally the Ahoms.

Rise and Fall of the Koch Kingdom

Danial could not rule his acquired territory for long. The local Bhuyans and other tribal chiefs unitedly fought the Muslims and killed Danial. Among these local heroes was one called Bisu who belonged to the Koch tribe of Goalpara[9] and who allied himself with the Bhuyans while fighting the Muslims. He successfully exterminated the latter and laid the foundation of the Koch kingdom in the first quarter of the 16th century.[10] The Koches were the co-brothers of the Mech-tribe of the Mongoloid stock and resembled, according to Minhas-ud-din-Siraj, the Mongoloid tribes of southern Siberia.[11] It appears from the *Darrang Raj Vamsavali*, a genealogical work of the Koch kings, that the Koches were adapted to tribal ways of living at the time and Bisu's family members largely cultivated cotton on the hills.[12] However, when he acquired power to control the surrounding situation, the Brahmanas, as pointed out by Gait, soon found him out and Hinduised him and his family, and called him Biswa Singha.[13] Out of gratitude he imported and settled a number of Brahmanas in his kingdom,[14] who not only accelerated the process of Hinduisation of the tribes in subsequent period, but might have also brought about some changes to the economic life of this tribal society.

Biswa Singha who ruled for about a quarter of a century (c. 1515-1540) had to acknowledge the overlordship of the Ahoms.[15] But his son and successor Malladeva *alias* Naranarayan (1540-1587 A.D.) threw away the yoke of Ahom suzerainty. Along with his younger brother and general Sukladhraj *alias* Chilarai, Naranarayan conquered the entire north eastern part of India including the kingdom of the Ahoms.[16] He also fought against the Karrani rulers of Bengal and helped Akbar to dethrone its rebellious ruler Daud about the year 1574 A.D.[17] But this glorious period of the Koch rule was short lived. After the death of Chilarai, the kingdom became divided into two parts—the region to the east of the river Sonkosh formed the kingdom of Koch-Hajo or Kamrup and that to the west formed the kingdom of Koch-Behar.[18] This happened in 1581 A.D.[19] This division of the kingdom leading to perpetual clashes between the two ruling houses in subsequent period, greatly weakened the Koches. The result was that not long afterwards the Koch-Behar kingdom was brought completely under the domination of the Mughals.[20] This event had badly affected the fate of the eastern one. The Mughal governors of Bengal after taking their rearguard in Koch-Behar, began to invade Koch-Hajo in subsequent period. During the reign of Jahangir (1605-1627 A.D.) its ruler Parikshitnarayan was captured and brought to Delhi, where he succumbed to death in 1613 A.D. and his son Balinarayan fled to the Ahom kingdom seeking protection.[21] As a matter of fact, this part of the kingdom was annexed to the Mughal empire[22] and the Mughal system of administration was then introduced there.

The occupation of Kamrup by the Mughals opened the way for direct clashes between the Mughals and the Ahoms. Prince Balinarayan who fled to the Ahom territory in 1613 A.D. as mentioned above, was established by the Ahom king Pratap Singha (1603-1641 A.D.) as a tributary king (Raja) in the region of Darrang between the Mughal and the Ahom territories[23] who formed an Ahom-Koch allied force to fight the Mughals. During the years that followed till at least the eighties of that century Ahom-Mughal wars continued almost unabated in course of which the Mughals gradually retraced so that during the times of the subsequent Ahom kings Udayaditya Singha (1669-1671) and Gadadhar Singha (1681-1696) they were further pushed back as far west as to the river Manah, which since then remained the permanent boundary of the Ahoms in the west till the occupation of the state by the British in the early part of the 19th century. This is how a greater part of the original Koch kingdom with both Koch and Mughal patterns of administration fell to the hands of the Ahoms and formed a part of their kingdom.

The expansion of the kingdom of Assam to the west as far as the river Manah, contributed in many ways to the growth of agrarian economy of the state. The plains of Kamrup, according to the contemporary records, produced varieties of fruits. Shihabuddin Talish, the court historian of emperor Aurangzeb, records that "Koch-Behar (and as such the rest of the kingdom) is well-known for its excellent water, mildness of climate, its fresh vegetation and flowers. Oranges are plentiful, as also other fruits and vegetables".[24] The *Alamgirnama* also speaks of the products of the Brahmaputra valley.[25] Varieties of rice like *ahu, bao*, and *sali* were grown by various sections of the people. The *Guru Charits* or the biographies of the Vaishnava saints of medieval Assam frequently mention such crops as grown in Kamrup regions. The kingdom was largely populated by the Garo and the Bodo Kachari tribes who lived in the submontane areas. Among them the use of hoe was much popular and *ahu* and *bao* varieties of rice were also largely cultivated.[26] Some of these tribes knew also the techniques of wet rice culture and irrigation. Even the Bhuyans, of whom we shall discuss later on, were much adapted to *ahu* and *bao*, i.e., sown broad-cast type of agriculture.[27] But in the plains region, *sali* or transplanted variety of rice was also grown as evidenced by the *Guru Charits* although its production was not large. The *buranjis*, i.e., medieval Assamese chronicles, record that it was from the Koch kingdom that a number of professional classes of people were imported to Assam.[28] This led to the growth of certain professions in Assam in subsequent periods. The fact that large number of coins were issued by the Koch kings and money economy was strictly maintained by the Mughals also proves that the economy was gradually monetised there. It is found that kings of Assam issued coins in large numbers only after their expansion towards the west. So the extension of the kingdom of Assam over a major part of the Koch territory had important effects on the economy of Assam in many ways. As we shall see later both the Mughal and the Koch systems of revenue administration were retained there even by the Ahom rulers of Assam to suit the circumstances.

Rise and Fall of the Chutiyas

Among the powerful tribal states formed in the period following the fall of the kingdom of Kamarupa was that of the Chutiyas, who belonged to the Mongoloid stock bearing close affinity with the Bodo Kacharis of Assam.[29] Their kingdom called Sadiya[30] extended in the north over the entire region from the river Sisi[31] in the west to the Brahmaputra in the east. The hills and the river Buri Dihing formed its northern and southern boundaries respectively.[32] Thus the Chutiya territory extended over almost the entire region of present districts of Lakhimpur, Dhemaji, Tinsukia and some parts of Dibrugarh. This indicates that over a major plains area now well-known for *sali* cultivation, this kingdom extended.

The early history of the Chutiyas can not be definitely known. The *Deodhai Asam Buranji* mentions that one Birpal *alias* Gayapal was the founder of the kingdom of the Chutiyas and had his capital at Sonagiri hills.[33] (i.e., near the source of the Sovansiri river). After him, his son and successor Gaurinarayan *alias* Ratnadhvajpal shifted his capital city to Ratanpur in the plains on the north bank of the Brahmaputra.[34] According to tradition he had finally built his capital at Sindhukhetra, which subsequently became known as Sadiya, which remained the head-quarter of the Chutiya kingdom till the last.[35] Extensive remains of buildings and fortifications built during the rule of the Chutiyas near about Sadiya still point to the importance of the region in the past. Thus the Chutiyas who were originally living in the hills, gradually came down to the plains and formed a powerful state. The transfer of this habitat brought about significant changes to their ways of living. They were very soon Hinduised and their kingdom also considerably prospered. The chronicles mention that there were people of Daivajna, Tanti, Sonari, Brahman, Sutar and Kamar castes in the Chutiya kingdom[36] at the time of its annexation by the Ahoms, indicating that by that time caste system became prevalent in their society. This had led to changes in their agrarian life as well.

The growing power of the Chutiyas, however, did not last long. After the death of Ratnadhvajpal the strength of the Chutiyas gradually declined. In the meantime the Ahoms to the south west of the Chutiyas established a powerful kingdom. Naturally the two states frequently came into clashes which, according to chronicles began since the beginning of the 14th century.[37] But the Chutiyas finally succumbed to the Ahom pressure and their king Suhungmung *alias* Dihingia Raja (1497-1539) completely subjugated the Chutiyas and permanently annexed their kingdom to that of the Ahoms in the year 1523 A.D.[38] Suhungmung then deported the Chutiya nobles and the prince Sadhaknarayan to his kingdom and established the later at Darrang with grants of land and labourers.[39] He also appointed an officer called *Sadiyakhowa Gohain* to govern the Sadiya territory.[40]

The annexation of the Chutiya territory by the Ahoms provided the latter with a fertile and heavy rainfall area in upper Assam. But then it is not definitely known as to the system of agriculture adopted by them. The fact that the Chutiyas were one of the earliest tribes of the valley to be Hinduised and to form a state, may point to their surplus economy. The growth of a number of professions among the people of this kingdom like *Tanti* (weaver), *Kahar* (bell metal worker), *Sonari* (goldsmith) as mentioned earlier, indicates the growth of some rural industries among the Chutiyas. The possible existence of a number of Bhuyans within the Chutiya territory[41] might have led to the growth of some advanced agriculture and trade. Besides, the Chutiya kingdom consisted of a vast plain level and fertile territory which provided for the Ahoms possibility of easy extension of wet rice culture in the region.

The Kachari Kingdom and Its Annexation

The Kacharis, a Bodo tribe of the Mongoloid stock, were one of the aborigins of the Brahmaputra valley.[42] In the time of the coming of the Ahoms in the 13th century they also had an organised state apparatus. Their kingdom was then extending over the regions between the rivers Dikhow in the east and the Kapili in the west in the south bank of the Brahmaputra and it extended to the Dhansiri valley including the areas which at present forms the North Cachar district of Assam.[43] The Ahom kingdom thus came to be existed on the immediate east of the Kacharis so that the latter soon had to bear the brunt of the westward expansion of the Ahoms. The Ahom king Suteupha (1268-1291 A.D.), son and successor of Sukapha (1228-1268) first expanded his kingdom in this direction by occupying a part of the Kachari possessions upto the river Namdang by purseing a policy of strategam.[44] It is stated in the chronicles that during the first half of the 16th century the Kacharis had to accept Ahom vassalage and were compelled to cede to the Ahom king Suhungmung not only a major part of their territory in the Dhansiri valley,[45] but also those territories beyond it as far as the Kalang in Nagaon.[46] The Ahom king then annexed the entire region into his kingdom and appointed an officer called *Marangikhowa Gohain* with his headquarter at Marangi to administer the newly acquired territory.[47]

The annexation of a part of the Kachari kingdom and its population contributed largely to the extension of Ahom agricultural lands in the plains of the Dhansiri valley covering parts of both Golaghat and Nagaon districts of present day Assam. The *Deodhai Asam Buranji* mentions how the Kacharis used to irrigate their fields by blocking small rivers.[48] Some of the Persian chronicles of medieval period also mention the technique of their making dams across the hill streams to bring water to the fields.[49] Even the early British records speak of the tribes as accustomed to practices of making dams and irrigating their fields.[50]

The Ahom rulers were therefore benefitted by the acquisition of the Kachari territories in two ways in the process of agricultural expansion—by way of acquisition of lands and by way of getting an already advanced class of cultivators. It should, however, be mentioned that even though the Kacharis were acquainted with irrigation, plough had probably very limited use among them. For almost all the early British records show that as late as the 19th century use of hoe among them was no less popular than that of plough.[51]

The Bhuyans and Their Subjugation

During the period under review, a class of land-lords called 'Bhuyans'[52] was ruling, at times independently and at times under some sovereign powers, in the plains of the Brahmaputra valley, particularly in the region now covered by the districts of Nagaon, Lakhimpur, Darrang and Kamrup. They were a land owning class, who by taking advantage of the fall of the early kingdom of Kamarupa rose to power in different areas. The princelets of the former ruling dynasties, erstwhile tributary rulers and tribal chiefs formed the class of the Bhuyans. Beginning from the 13th century to the 15th century, the period of three hundred years witnessed strong Bhuyan domination in the middle of the Brahmaputra valley.

The Bhuyans were chiefly instrumental in dethroning Danial, the Muslim governor of Kamata, established by Ala-ud-din Hussain Shah in 1498 A.D. Soon after, however, Koch king Biswa Singha, established his sovereignty over the Bhuyans and compelled them to offer allegience to him. The *Darrang Raj Vamsavali* mentions how Biswa Singha defeated the Bhuyans one after another.[53] But he failed to root out the Bhuyan rule altogether and throughout the Koch rule in the 16th and the 17th centuries the Bhuyan institution remained as a state within a state.[54]

The Bhuyans outside the Koch territories proved a stumbling block to the expanding Ahom kingdom in the north and the west. Suhungmung Dihingia Raja had to wage long warfare to subjugate them.[55] King Pratap Singha also had to deal with some of the Bhuyans and succeeded in bringing them under his control.[56]

The Bhuyans played significant roles in the economy of medieval Assam. They reclaimed new lands and brought these under cultivation. They had their own armed forces[57] and exercised every power in their own territory. They had also man-power enough to cultivate extensive lands. In fact, they resembled the autonomous chiefs of Mughal India who originated independently of the existing imperial power.[58] The state claimed from them only a share of their total collection, not the land revenue as such.[59] Some of the Bhuyans were expert in trade and commerce and were responsible in carrying on trade between upper and lower Assam.[60] The *Katha Guru Charit* mentions how the Bhuyans of Nagaon collectively built embankments to facilitate wet rice cultivation.[61] The same work, as well as the *Yogini Tantra* make mention of varieties of *sali* and *ahu* paddy[62] grown in the area mostly dominated by the Bhuyans. It was through their initiative that the use of plough and wet rice culture became spread amongst some of the neighbouring tribes of contemporary Assam. Expert artisans by themselves, many Bhuyans were engaged in weaving and other industrial works under the Koches.[63] The occupation of the Bhuyan territories, therefore, naturally helped the Ahom rulers to extend agricultural pursuits among their population.

The Ahom Migration and Formation of the State of Assam

The most significant role in Assam in the growth of its civilization, was played by the Ahoms, a section of the Tai (speaking group) of the Mongoloid race of south-east Asian regions, who originally dwelt in Mongolia and in the south-western part of China in the south of Yang-tse-Kiang.[64] In course of time, as a result of their conflicts with the Chinese people,[65] some of them had migrated as far as Yunan and established there a number of principalities.[66] During the sixth and the subsequent centuries they expanded over and occupied the whole of the Sheuli valley down to the Irrawaddy in upper Burma and formed there a number of Tai states, such as, Hsen-wi, Mong-mit, Bhamo etc.[67]

According to P. Gogoi, it was from the Tai state of Mong-mit that the Ahoms under the leadership of Sukapha came to Assam.[68] But there have been a number of contradictions as to the place from where Sukapha and his people came to Assam. S.K. Bhuyan on the basis of the Tai-Ahom and the Assamese chronicles states that it was Mao-Lung within the kingdom of Mung-Mao, a Shan state in upper Burma[69] from where Sukapha came. In Gait's history[70] also Sukapha's original home land is mentioned as Mao-Lung. In the Ahom chronicles it is variously mentioned as Mao-Lung, Mong-Mao Mong-Mao-Lung etc.[71] According to J.N. Phukan these names stand for the same state and belong to the 'Big Mao' or 'Big Tai' branch of the Tais living in south-western Yunan and upper Burma; but its proper boundary is difficult to define.[72] On the basis of the similarity of the Tai scripts used by the Ahoms of Assam and the Tais of south-western Yunan (The Dehong Prefecture), he opines that the state was somewhere in Yunan.[73] According to E.R. Leach, existence of this state of Mao-Lung or Mong-Mao may be placed somewhere between the Nam-Mao and Nam-Wan rivers some fifty miles to the east of Bhamo in Burma.[74] It is, however, not feasible to locate the people within such limited boundaries. From the extent of the spread of the same people over a wide range of territories, we may conjecture that the Tais or the 'Shan proper' to which Sukapha and his people belonged, might have spread from western Yunan to upper Burma and it was from the nearest station that they came to the Brahmaputra valley.

Sukapha was accompanied by a large number of followers. The chronicles mention various numbers between 480 to 9000.[75] The object of founding a new kingdom in a hithertho unconquered land makes it convincingly clear that Sukapha was accompanied by a large number of followers most of whom were men. According to P. Gogoi, from the manner of description given in the *buranjis*, it appears that, barring other followers, the regular fighting force of Sukapha was nine thousand strong, as the contingent had three thousand *matang* or cooking vessels, made of copper, each for a squad of three men.[76] Besides, Sukapha had with him 300 horses and two elephants leaving aside a number of swords called *hengdang*.[77]

Sukapha moved towards Patkai amidst dense jungles inhabited by ferocious tribes. But he conquered the region around the Daikham mountains south of Patkai and entered the Naga territory wherefrom he crossed the river Khamjang and reached the Nangyang lake.[78] He defeated the Nagas there and made it a province called Khamjang and appointed one Kang-Khru-Mong as its governor.[79] Sukapha thenceforth marched towards the Patkai range and conquered the Naga territories. It is mentioned that so severely he had dealt with the Nagas that those unconquered voluntarily submitted to him.[80] Then he reached Numruk (Namrup) and Tipam and entered the valley of the Brahmaputra,[81] in 1228 A.D. The whole course of action since his leaving Mao-Lung took a period of thirteen years.

Sukapha then moved from one place to another in search of a suitable place for his capital. The Tai-Ahom and the Assamese chronicles mention that in this way he moved to Abhaypur (Munklang Chekhru) in 1236 A.D., then to Habung, then went down-stream of the Brahmaputra to the mouth of the Dikhow river and thenceforth to Ligirigaon. In 1246 he reached Simaluguri and stayed there for some years and finally he reached Charaideo in 1253 A.D.,[82] and established his headquarter there.

At the time of Sukapha's arrival, the land was inhabited by some tribes who were in their primitive stage of production. Among them, the Marans were occupying the region between the Burhi Dihing and the Dichang river, the Barahis were inhabiting the territory between the Dichang and the Dikhow—both on the south bank of the Brahmaputra. These regions were contiguous to Charaideo. As mentioned in the chronicles, Sukapha's men estimated the total population of both the tribes at about four thousand souls.[83] The chief of the Marans was Badaucha and that of the Barahis was Thakumtha, both of whom Sukapha brought under his control by friendly means.[84] The chronicles mention that since then the Marans and the Barahis began supplying the Ahom kings with articles of their daily need.[85] Sukapha cemented his friendly relations with these local tribes by marrying their daughters and appointing their men as cooks (*changmai*) in the royal kitchen.[86] According to the *Deodhai Asam Buranji* since that time, there was an admixture of blood, and children were of mixed origin, as the Ahoms had not brought their wives when they first came and as they accepted wives only when they came here.[87]

Thus by the middle of the 13th century, the Ahoms established a small kingdom centering round Charaideo, in performing which, they received all cooperation from the local tribes, who also belonged to the same stock of the Mongoloid racial group as that of the Ahoms. But unlike these tribes who were then in a primitive stage of cultivation, the Ahoms were already having an improved culture and economy based on developed agricultural systems. This can be inferred from the fact that with a view to collecting revenue, Sukapha had already made arrangements for development of agriculture in Khamjang before he had crossed the Patkai hills.[88] Besides, he made administrative arrangements of the territories he conquered on his way and kept his people in those places in order to engage themselves in agricultural pursuits.[89] He himself changed places for a number of times due to flood, which was a hindrance for agriculture, before he finally settled down at Charaideo.[90] That the Tais were great agriculturists is hinted in the annals of the Chinese, who in their farm were using animals like buffaloes and oxen, and were developing irrigation facilities for the purpose of cultivation.[91] The Tai legend about their origin contain distinct reference to their objective of pursuing agricultural activities behind their establishment of a kingdom on the earth.[92] From such references, it can be presumed that at the time of coming of the Ahoms, the local tribes, who were mostly habituated to shifting method of agriculture, were keeping vast areas lying fallow.[93] The Assamese chronicle translated by J.P. Wade confirms such presumption which mentions that at the time of Sukapha's arrival in the easternmost part of the Brahmaputra valley, the region was wild and uncultivated.[94] It therefore appears that the Ahoms had a mission to fulfill in Assam—the mission of making it green with agricultural pursuits, and Sukapha as the founder of their kingdom laid the first step in that direction.

Sukapha died in the year 1268 A.D. after a reign of forty years. Before his death, he not only laid the foundation of a kingdom on a strong and permanent footing, but also started the process of bringing about a radical change in its material life through the introduction and expansion of wet-rice culture. His long march since his crossing over of the Patkai hills to the establishment of his capital city at Charaideo, was an experiment of testing soil and its fertility to suit this method of cultivation. The chronicles mention that Sukapha selected the place around Charaideo because of the fact that he found the water of the river Santak flowing nearby to have contained more alluvium than any other river.[95] He is also said to have established three big farms in the suburbs of his capital for growing different varieties of *sali* rice, known as the *Barakhowakhat*, the *Engerakhat*, and the *Gachikala khat*[96] with the help of local recruits. The chronicles also mention that he appointed two of his accompanying senior nobles as ministers of the kingdom, called *Chao Phrang Mung* and *Chao Thao Lung*,[97] which subsequently came to be known as *Buragohain* and *Bargohain* respectively. It was again he, who introduced the process of recording the political events of the state which came to be known as the *buranji*.[98]

Sukapha's successor Suteupha ruled for about fourteen years. During his reign the kingdom extended in the west as far as the Dikhow river. He was succeeded by his son Subinpha in 1291 A.D. during whose rule the system of assignment of men to the state officers in lieu of pay was introduced. The entire population of the kingdom was divided and put under the king and the two great officers—*Buragohain* and *Bargohain*.[99] They were also instructed to supply the king with certain amount of their produce for the maintenance of the king.[100] Besides, the people of the local origin who were subdued were distributed among the nobles to serve as domestic workers.[101] During the reign of Subinpha's successor Sukhangpha (1293-1332) the Ahom kingdom grew in power and its influence was felt as far west as the kingdom of Kamata. The chronicles mention that the then Kamata king (possibly) Pratapdhvaj (c. 1305-1325 A.D.) entered into a matrimonial relation with this Ahom king by marrying his daughter Rajani to him.[102] The increasing power of the Ahoms was not liked by the neighbouring Chutiyas. This resulted in a series of clashes between the two powers in the years that followed.[103]

The last years of the 14th century saw the rule of a king who introduced a number of new elements to the Ahom state. He was Sundangpha *alias* Bamuni Konwar (1397-1407) who was said to have been born and brought up in a Brahmana family of Habung.[104] He introduced certain Brahmanical elements in the royal household and appointed Brahmana counsellors in the court.[105] This had created much resentment among the orthodox Ahoms, notably the Tipamiyas who not only revolted against the king[106] but also compelled him to introduce at the same time certain new Ahom rites in the royal court.[107]

Sudangpha's successors were capable of maintaining their territorial integrity but could not achieve much expansion. As a result, till the end of the 15th century, the limits of the Ahom kingdom remained confined only to the south bank of the Brahmaputra. One of their kings, called Susenpha (1439-1444 A.D.) first built the Nagsankar temple in the north and this act of the king, as put by a historian, gained him the goodwill of the Aryanised Mongoloid tribes of the region, who helped him in his subsequent conquests of Habung and Panbari[108] and in extension of agriculture in the north bank.

The Ahom kingdom was firmly established and markedly extended by Suhungmung *alias* Dihingia Raja (1497-1539). He was the first Ahom ruler to assume a Hindu name along with one in Tai, and was called *Swarganarayan* or *Swargadev*,[109] which thenceforth became the title of the Ahom king. He vigorously followed an expansionist policy in all directions. It was during his reign that the Chutiyas were completely subjugated and their kingdom annexed to his dominion. This has been discussed earlier. To complete this work of subjugation, he established three hundred Ahoms of the *Gharphalia* clan at Sadiya, the capital of the Chutiyas, and another contingent of the same clan was established by the side of the river Dihing.[110] The establishment of the Ahom families in Sadiya may be taken as a turning point in changing the pattern of agriculture of the region. It should be mentioned that this king also built a big embankment on the bank of the Dihing river to prevent its flood and established his headquarter there.[111] During his reign the Kacharis were also expelled from the plains region. Besides, he subjugated the Bhuyans, made the Koch king his tributary and concluded alliances with the rulers of Orissa and Manipur.[112]

The growing power of the Ahoms under Suhungmung attracted the attention of the Muslim rulers of Bengal. From the records of the *buranjis* it is known that there were a number of invasions led by the Sultans of Bengal during the first half of the 16th century,[113] in one of which about 2,500 of their soldiers were captured, who were subsequently settled in Assam. They were later on formed into a *khel* (guild) known as the *Mariya* and were engaged to work on brass.[114]

The chronicles record that Suhungmung's successor Suklenmung *alias* Gargaya Raja (1539-1552 A.D.) engaged both the Ahoms and the subjugated local tribes to collectively open lands on the bank of the river Dikhow for agricultural pursuits and personally supervised their works.[115] The process of emergence of the Ahom kingdom as the only major power of the region with its strong agrarian economy was, however, not complete till the time of king Susengpha *alias* Pratap Singha (1603-1641 A.D.), who not only fought the Mughals for almost four decades, but also formed a concrete civil society in Assam. It was during his reign that the tribes of the Brahmaputra valley had gradually begun accepting the Ahom ruler as their overlord. The *buranjis* mention that during the Ahom Mughal wars a number of tribal chiefs came forward to cooperate with the Ahoms against the invaders.[116] The role of Pratap Singha is significant in more than one way in giving Assam some permanent socio-economic structures. It was during his rule that Momai Tamuli Barbarua established model villages and introduced the *paik* system in a systematic way[117] which remained till the end the basic character of the agrarian life of the people of Assam under the Ahoms. It was through the *paik* system that lands were brought under regular and settled cultivation by their equal distribution among the *paik*-workers, and jungle lands were also cleared for gradual expansion of wet rice cultivation. Besides, the king built a number of embankment-cum-roads with *paik* labourers which were largely in use to control flood and protect the agricultural fields.[118] During his reign, the Neo-Vaishnavite movement became spread in Assam. The organisational character of this sect[119] and the *paik* system both had great impact on the life of the people as a result of which the tribal character of the society got underminded and a form of feudalism grew up.

The Ahoms had to wage protracted wars against the Mughals for at least another four decades even after Pratap Singha to establish permanently their western boundary and make their kingdom the most dominant one in the entire north-east. During this period the rulers of Assam tried their best to improve the economy of the state by various ways to successfully withstand the Mughal invaders. The chronicles mention that king Chakradhvaj Singha (1663-1669 A.D.) was advised by his Prime Minister Atan Buragohain to increase production to strengthen the state's economy before launching any war against the Mughals and achieve success in it.[120] In fact, the king personally supervised all works of production including rice and its provisions of supply, and himself taught the *paiks* who were found deficient in their works.[121] It should be noted that the sound and stable economy of the state played the most important role in enabling the Ahoms to resist the attacks by the mighty Mughals. In the final stage after the battle of Saraighat in 1671, Hadira opposite to Goalpara became the frontier outpost in the west of the Ahom kingdom. This *status quo* was broken for a short period during 1679 A.D., but the same was restored by king Gadadhar Singha who by defeating the Mughals in the battle of Itakhuli reoccupied the region upto the river Mahah in 1682 and made it a permanent part of the Ahom kingdom.[122] Thus the part of the territory of the former Koch kingdom which had fallen to the hands of the Mughals in 1612 A.D. became now a part of the Ahom kingdom and remained with it till it was taken over by the British in the early part of the 19th century.

Thus a small state formed in the extreme south-east corner of the Brahmaputra valley in the early part of the 13th century became the single largest state of Assam covering almost the entire region of the valley by the last part of the 17th century. The political expansion was accompanied by the growth of population of diverse ethnic groups, professions and cultures. At the background of the formation and growth of the state was its economy which was largely based on its agriculture. The Ahom rulers from the very beginning adopted the policy of extension of agriculture by way of opening new lands, elevating the fields, building embankments for protection of cultivated fields from floods, and above all, by collecting and engaging every adult male in agricultural pursuits. In the subsequent chapters other aspect of their agrarian economy are being discussed.

REFERENCES

1. For the word 'Asam' see E.A. Gait, *A History of Assam*, 2nd edn. Reprint, Guwahati, 1981, pp. 245f (henceforth abbreviated as *HAG*); B. Powell, *The Indian Village Community*, First Indian Print, Delhi, 1972, p. 135; S.K. Chatterji, *The Place of Assam in the History and Civilization of India*, 2nd Print, Gauhati University, 1970, p. 42; B. Kakati, *Assamese: Its Formation and Development*, Guwahati, 1941, pp. 2f; J.N. Phukan, 'Presidential Address' in the *Proceedings of North-East India History Association*, 1989, pp. 6f.

2. The *Deodhai Asam Buranji*, (ed), S.K. Bhuyan, 2nd edn, Guwahati, 1962, p. 100 (henceforth abbreviated as *DAB*); *Satsari Asam Buranji* (ed), S.K. Bhuyan, 2nd edn, Guwahati, 1964, p. 5 (henceforth *SAB*).
3. The Bhuyans were land-lords of various ethnic origins who ruled either as independent chiefs or as feudatories under some powerful overlords. See K.L. Barua, *Early History of Kamarupa*, 2nd edn., Guwahati, 1966, pp. 184ff; D. Nath, 'The Bhuyan Land-Lords and Their Role in the Economy of Medieval Assam', *India Past and Present*, Vol. III, No. 2, Bombay, 1986, p. 245.
4. See H.C. Raychaudhuri, *Political History of Ancient India*, 7th edn, Reprint, Calcutta, 1972, pp. 166ff.
5. *DAB*, p. 133; *Katha Guru Charit* (ed.), U.C. Lekharu, 15th edn, Nalbari, 1987, pp. 85, 108, 109 (henceforth *KGC*); *Darrang Raj Vamsavali* (ed), N.C. Sarma, Pathsala, 1973, pp. 341, 364 (henceforth *DRV*).
6. See K.L. Barua, *op. cit*, pp. 160ff.
7. *Ibid.*, p. 170f.
8. Ghulam Hussain Salim, *Riyaz-us-Salatin* (tr.), Abdus Salam, Reprint, Delhi, 1975, pp. 132ff; also see K.L. Barua, *op. cit*, p. 174.
9. *DRV*, v. 51; *Kamrupar Buranji*, (ed.), S.K. Bhuyan, Guwahati, 2nd edn, 1958, p. 10; *DAB*, pp. 173f; *HAG*, p. 48; D. Nath, *History of the Koch Kingdom, 1515-1615*, New Delhi, 1989, p. 16.
10. *HAG*, p. 48; K.L. Barua, *op. cit*, p. 190; Nath, *op. cit*, p. 26.
11. *Tabaqat-i-Nasiri*, (tr.), H.G. Raverty, Reprint, New Delhi, 1970, p. 560.
12. vv. 53-54, also see *Kamrupar Buranji*, p. 10.
13. *HAG*, pp. 48f.
14. *Ibid.*, p. 49; also in one MS chronicle in Assamese preserved in the *DHAS*, Guwahati.
15. *DAB*, p. 107; *Ahom Buranji*, (tr.), G.C. Barua, Reprint, Guwahati, 1985, p. 77 (henceforth *AB*).
16. *DRV*, vv. 317-488; *HAG*, pp. 51ff; Nath, *op. cit*, pp. 50ff.
17. *Riyaz*, p. 120; *Akbarnama*, Vol. 3 (tr.), H. Beveridge, (*Bibliotheaca Indica* series, Calcutta), Reprint, Delhi, 1973, p. 1068; also see S.N. Bhattacharya, *The Mughal North East Frontier Policy*, Calcutta, 1929, pp. 97f; Nath, *op. cit*, pp. 72ff.
18. *DRV*, vv. 648-49; *KGC*, p. 615.
19. *HAG*, p. 56.
20. *Akbarnama*, Vol. 3, pp. 168, 1081ff; *Bharistan-i-Ghaybi* by Mirza Nathan Ala-ud-din Isphani, (tr.), M.I. Bora, Vol. I, Guwahati, 1936, p. 40.
21. *DAB*, p. 107; *AB*, p. 77; J.P. Wade (tr.), *An Account of Assam*, North Lakhimpur, 1927, p. 221.
22. *Baharistan*, I, p. 272; *Kamrupar Buranji*, p. 131; *HAG*, p. 63; Wade, *op. cit*, pp. 220f.
23. *DAB*, p. 109; Wade, *op. cit*, p. 220; Nath, *op. cit*, pp. 130f.
24. *Fathiya-i-Ibriya* by Shihabuddin Talish, (tr.), Sir J.N. Sarkar in the *Journal of Bihar and Orissa Research Society*, 1915, Vol. I, Pt. II; H. Blochmann, 'Koch Behar, Koch Hajo and Assam in 16th and 17th Centuries according to the *Akbarnama*, the *Padishnamah* and the *Fathiya-i-Ibriya*, *JASB*, 1872, Vol. LXI, Pt. I, p. 66.
25. (Tr.) Henry Vansittart in the *Asiatic Researches*, Vol. II, 2nd edn, New Delhi, 1979, pp. 130f.
26. John Butler, *A Sketch of Assam with Some Accounts of the Hill Tribes*, London, 1847, pp. 21f.

27. *KGC*, pp. 56, 59, 63, 64, 76.
28. *Asam Buranji* (obtained from the family of Sukumar Mahanta), (ed.), S.K. Bhuyan, 3rd edn, Guwahati, 1988, p. 38 (henceforth *ABSM*); *Asam Buranji* by Harakanta Barua Sadaramin, (ed.), S.K. Bhuyan, 2nd edn, Guwahati, 1962, p. 36 (henceforth *ABHB*); *Asam Buranji Sar*, by Kasinath Tamuli Phukan, (ed.), P.C. Choudhury, 3rd Print, Guwahati, 1964, p. 24 (henceforth *ABS*).
29. S. Endle, *The Kacharis*, London, 1911, pp. 4ff.
30. S.L. Barua, *A Comprehensive History of Assam*, New Delhi, 1985, p. 183 (henceforth *CHAB*).
31. On the basis of two land-grants of *Saka* 1314 (1392 A.D.) and 1323 (1403 A.D.) S.L. Barua has contended that this kingdom extended to the west as far as Biswanath in present Sonitpur District of Assam. See *ibid.*, pp. 182f.
32. *Ibid.*, p. 182.
33. p. 193.
34. *Ibid.*, p. 194.
35. Cf. *CHAB*, p. 103.
36. *DAB*, pp. 126, 194; *SAB*, p. 14.
37. *DAB*, p. 11; *CHAB*, p. 185.
38. *Ibid.*, p. 200; *SAB*, pp. 66ff; *ABS*, pp. 17f; *ABHB*, p. 24.
39. The chronicle mentions that Suhungmung issued a copper plate granting the Chutiya prince land and labourers with the provision of getting as tribute from him eighty rupees and forty-blankets annually in return. See *DAB*, p. 200.
40. *Ibid.*, p. 19; *SAB*, pp. 60f; *ABHB*, p. 24.
41. One of the legends of the origin of the Bhuyans associate them with Lakhimpur which was within the Chutiya territory; see the *Bhuyanr Puthi (Adicharit)* cf. *CHAB*, p. 194.
42. S. Endle, *op. cit*, pp. 4f; *HAG*, p. 247.
43. *HAG*, p. 248; *SAB*, pp. 7f; The Kacharis were the most numerous aboriginal tribe of the Brahmaputra valley. Prior to their formation of a powerful state in the south of the Brahmaputra including Cachar, they had formed also a state in Sadiya which was bounded, according to the editor of the *Kachari Buranji* (ed), S.K. Bhuyan, Guwahati, 3rd edn, 1984, Introduction, p. VII, by Sadiya in the east, the river Dikhow on the west, the Dihing on the north and the Dilli river on the south.
44. *DAB*, pp. 102f; *SAB*, pp. 7f.
45. *Ibid.*, pp. 26ff; 35ff; *ABSM*, pp. 11ff; *AB*, p. 64.
46. *ABS*, p. 18; *ABHB*, p. 25.
47. *Ibid.*, p. 27; L. Devi, *Ahom Tribal Relations*, Guwahati, 1968, p. 84.
48. *DAB*, p. 100.
49. *Tabaqat-i-Nasiri*, p. 565.
50. Butler, *op. cit*, p. 23.
51. Francis Hamilton, *An Account of Assam*, (ed), S.K. Bhuyan, Guwahati, 1940, 2nd edn, 1963 (henceforth *AAH*), p. 73; E.T. Dalton, *Tribal History of Eastern India*, 2nd print, New Delhi, 1974, p. 81; Boden Powell, *Land Systems of British India*, Vol. 3, Delhi, 1974, pp. 41ff.
52. The term 'Bhuyan' has been derived from the Sanskrit word *Bhaumik*, meaning land-lord which is equivalent to the Persian word *Zaminder*. See J. Wise 'on the Barah Bhuyans of Bengal' in *JASB*, 1974, Pt. I, No. III, p. 197; *HAG*, p. 38.
53. vv. 90-122; also see Wade, *op. cit*, p. 187.
54. *India Past and Present*, Vol. III, No. 2, pp. 245f.

55. *ABS*, p. 18; *ABSM*, p. 11; *SAB*, pp. 16f.
56. *SAB*, p. 78.
57. *KGC*, pp. 48f.
58. I. Habib, *The Agrarian System of Mughal India (1556-1707)*, Bombay etc. 1963, p. 182.
59. *Ibid.*
60. *KGC*, p. 86; *India Past and Present*, Vol. III, No. 2, p. 248.
61. *KGC*, pp. 11f, 38.
62. *Ibid.*, p. 26, 56, 75; *Yogini Tantram* (tr.) Swami S. Saraswati, 1385 Bengali *Saka* (1979), vv. 295-97.
63. *KGC*, p. 267.
64. P. Gogoi, *The Tai and The Tai Kingdoms with Fuller Treatment of the Tai Ahom Kingdom in the Brahmaputra Valley*, Guwahati, 1968, pp. 32, 102; W.A.R. Wood, *A History of Siam*, 2nd edn, Bangkok, p. 32; Phukan, *loc. cit*, p. 6.
65. Gogoi, *op. cit*, pp. 32ff; J.N. Phukan, 'The Tai Ahom Power in Assam' in *The Comprehensive History of Assam*, Vol. II (ed), H.K. Barpujari, Guwahati, 1992, p. 50.
66. Gogoi, *op. cit*, p. 103.
67. *Ibid.*
68. *Ibid.*, p. 253.
69. S.K. Bhuyan, *Anglo-Assamese Relations*, Guwahati, 1949, p. 2.
70. *HAG*, p. 77.
71. *AB*, pp. 4, 42f; *DAB*, pp. 4f; To the Manipuris it was known as 'Pong'; see *HAG*, p. 70.
72. *Loc. cit*, p. 4.
73. *The Comprehensive History of Assam*, II, p. 51.
74. E.R. Leach, *Political Systems of Highland Burma*, 1954; cf. Phukan, *loc. cit.*, p. 5.
75. See *AB*, p. 44; *DAB*, p. 6; the *Purani Asam Buranji* (ed), H.C. Goswami, Guwahati, 1977, p. 10 (henceforth *PAB*); *ABHB*, p. 11; *ABS*, p. 9; *ABSM*, pp. 3, 5; *SAB*, p. 45. From the variations in the number it appears that the number of the followers who accompanied Sukapha from Mao-Lung, gradually increased on the way when additional groups joined the march.
76. *op. cit*, p. 257.
77. See *DAB*, p. 6; *ABHB*, pp. 4f.
78. P. Gogoi, *op. cit*, p. 259.
79. *DAB*, p. 7; *AB*, p. 45.
80. *Ibid.*, pp. 7f; *ibid.*; Gogoi, *op. cit*, p. 261.
81. *Ibid.*, p. 8; *ibid.*, pp. 45f.
82. *Ibid.*, p. 8; *ibid.*, p. 46; *SAB*, pp. 46f.
83. *Ibid.*, p. 100; *SAB*, p. 5.
84. *ABS*, p. 10; *DAB* mentions only one chief called Thamithuma who ruled over the Marans and the Barahis together. See p. 100.
85. *DAB*, p. 101, *SAB*, pp. 6f; *ABS*, p. 10; *ABHB*, p. 12.
86. *Ibid.*
87. p. 99, Some scholars, however, believe that Sukapha had with him three wives, and bands of village folks including women and children had joined the march. See P.K. Buragohain, *The Tai Sanskriti*, Guwahati, 1956, p. 81; Gogoi, *op. cit*, p. 258.
88. *ABHB*, pp. 11ff; *ABS*, p. 7.

89. *DAB*, 100; *SAB*, p. 3.
90. *ABGB*, p. 47; *DAB*, pp. 8f; *AB*, pp. 44f; *ABS*, pp. 7f.
91. T. Fitzsimmons (ed), *Thailand*, New Haven, 1956, pp. 59ff; A. Guha, *Medieval and Early Colonial Assam Society, Polity, Economy*, Calcutta, 1991, p. 63 (henceforth *MEAG*).
92. Legend goes that in a series of discussion in heaven among the gods, and goddesses about the feasibility of founding a kingdom on earth, Lengdon, the supreme God, is said to have stated in this way: "There are no kings in the countries under the sky. Large fields are lying fallow. These may be well cultivated. The people of up and down countries are in constant warfare with each other and whoever may get victory rules the countries for the time" see *AB*, p. 10.
93. *MEAG*, p. 63.
94. *An Account of Assam, op. cit*, p. 18.
95. *DAB*, p. 8.
96. *ABHB*, p. 3; *ABS*, p. 10.
97. *DAB*, pp. 6f.
98. *Ibid.*, p. 98; *SAB*, p. 3.
99. *Ibid.*, p. 103; *ibid.*, p. 8.
100. *Ibid.*, p. 103; *SAB*, p. 8.
101. *Ibid.*; *ibid.*
102. *Ibid.*, p. 10.
103. *HAG*, p. 83.
104. *ABS*, p. 14; *SAB*, p. 11; *AB*, p. 49; *DAB*, p. 106.
105. *SAB*, p. 11; *AB*, p. 49.
106. *DAB*, p. 11; *ibid.*, p. 49f.
107. Sundangpha was the first Ahom king to introduce the custom of *Singari Ghar Utha* (to climb the plate-form of the *Singari* wood) after accession to the throne; see *ABHB*, p. 19; *ABS*, pp. 14f; *CHAB*, p. 226.
108. *CHAB*, p. 227.
109. *ABHB*, p. 28; *HAG*, p. 86; Gogoi, *op. cit*, p. 283.
110. *DAB*, p. 19; *ABGB*, p. 57; *HAG*, p. 38.
111. *ABS*, p. 17; *DAB*, p. 16; *ABHB*, p. 23.
112. *ABSM*, pp. 20f; *SAB*, p. 69; *DAB*, p. 38; Wade, *op. cit*, p. 191.
113. *DAB*, pp. 22f, 28, 32; H. Barbaruah, *Ahomar Din*, Guwahati, 1981, p. 55 (henceforth *ADB*).
114. *ABS*, p. 20; *ABHB*, p. 28; *DAB*, p. 200.
115. *SAB*, p. 20. The work records that the non-Ahom workers (*Raj-mari-ana*) finding the works tough enough when became reluctant to continue, the king had to remind them that since the lands opened by them would be enjoyed by their descendants in future by paying revenue to the state, they should not be reluctant to open it.
116. *CHAB*, p. 253.
117. *DAB*, p. 130; *ABSM*, p. 63.
118. *SAB*, p. 76. The king himself accompanied the officers-in-charge of making roads-cum-embankments and directed the route for their construction.
119. M. Neog, *Sankaradeva and His Times*, Guwahati, 1965, pp. 360ff; *CHAB*, p. 254.
120. *PAB*, pp. 103f.
121. *Ibid.*, p. 104.
122. *ABSM*, p. 123; *SAB*, pp. 110f; *ABS*, p. 40; *HAG*, p. 159.

2

LAND SYSTEM

Classification of Lands

During the medieval period land and men were the major sources of revenue in Assam. Land was the source of every item of agricultural products as well as of metals, salt, fishery and forest and animal resources; and it was on these items that the income and existence of the state largely depended. At a time when there was no large scale industries to replace manual labour, men were the only means of all sorts of production. As a result, the rulers of Assam seriously undertook the business of accounting lands and labour so that there may not be any lapse in the utilisation of lands and available man-power.

The Ahoms who integrated almost the entire Brahmaputra valley by subjugating the local ruling tribes and formed it into a stable state, were an agricultural tribe and inherited the culture of productive rice from their original homeland.[1] From the legend which records the migration of the Ahoms, it appears that they moved to the Brahmaputra valley in search of cultivable lands. The terms *Mung-dun-Sun-Kham* and *Mung-ri-Mung-ram* associated with it implied lands of brighter prospects for agricultural pursuits.[2] The Ahom legends about the origin of the Tai people record that the gods in heaven sent the forefathers of the Ahoms to the earth because the heaven was gradually becoming more populated while at the same time the earth was lying vacant,[3] with large fields lying fallow which could be easily brought under cultivation.[4] As pointed out by Guha, thus "the migrant Ahoms believed that they had a mission to introducing better cultivation"[5] there (i.e., Assam). The fact that at the time of the Ahom migration to the Brahmaputra valley, almost the entire upper region of the valley was very thinly populated,[6] points to the significance of the legend. So the Ahoms since their migration to Assam began a planned expansion of agriculture in the region. This needed besides clearing, surveying and proper classifying of lands.

Land survey and measurement were not unknown to the people of ancient Assam. Land classification was taken as a measure of revenue administration in the kingdom of Pragjyotisha-Kamarupa several centuries before the coming of the Ahoms. This is confirmed by the epigraphs of the period recording donation of lands to the Brahmanas and the temples by the rulers of ancient Assam. These epigraphs mention usually three classes of land in use. These were *kshetra* or cultivable or cultivated lands, *khila* or arable waste lands and *vastubhu* including *vithi* or homestead lands.[7] While *vastubhu* or homestead lands included the housing plot, the flower, fruit or vegetable garden and the bamboo groves, the *khilas* or the waste lands included forests, ponds and marshy lands. This system of land classification did not become obsolate during the rule of the Ahoms. But even retaining this old division, they classified lands into minute details. This was due to the fact that the Ahom rulers were very keenly interested in bringing under cultivation whatever lands they could clear in the valley. To realise this objective, they took all measures to convert the uncultivable or unutilised lands under their control into first class cultivable land called *rupit* or *rowati* by recruiting personal service from the peasantry.[8]

The Ahom rulers classified lands according to their use. Their epigraphs mention cultivated, cultivable and uncultivated lands. Cultivated lands were of various classes or types—the foremost being *rupit*[9] or land under permanent cultivation where *sali* or wet rice was grown. *Rupit* lands were treated as the first class land because this class of land formed the major source of the state's total revenue collection. These lands were distributed among the cultivators in lieu of which the government recruited man-power for the use of the state, or collected cash revenue at the rate of Rs. 2 per *pura* (i.e., a unit of 4 *bighas* of land) when the cultivator failed to supply his labour to the state.[10] As *rupit* was the most substantial class of land, the government took much interest in increasing its extent. The second important category of lands suitable for cultivation was *pharingati*,[11] i.e., lands lying uncultivated, but which were gradually brought under permanent cultivation. It was a policy of the Ahom kings to encourage cultivators to clear jungle lands for cultivation of seasonal or temporary crops. Land thus cleared were not first class revenue paying ones. Until its surface was made level plain to make it suitable for *sali* cultivation, the cultivators had to grow for several years their winter crops like *ahu, bao*, different kinds of pulses, mustard seed etc., which were sown broad-cast. Such lands were therefore known also as *ahutali, baotali, sariyahtali* or sometimes as *maghuwatali*,[12] for those lands were low lying and were to be cultivated during winter in the month of *Magh* (middle of January to middle of February). Some of the *pharingati* lands were high lands, unsuitable for wet rice cultivation, which included *habimati* (lands covered by jungles or reeds) and *bori* or cultivable open high lands. The Ahom kings encouraged the peasants to cultivate in these lands also. As a result these lands were gradually brought under wet rice cultivation by the process of levelling the surface and providing them with small and cris-cross ridges so that it could contain rain water. The cultivators who opened such lands and brought them under temporary cultivation, were not required to pay to the state any revenue in terms of labour, but had to pay Re. 1 in cash per *pura* of land to the treasury of the state.[13] These were also known as *katani* or *obar* or *obarmati*.[14] In lower Assam this type of land was called *jamar ubar*, or land reserved for conversion in near future to wet rice cultivation.[15] Uncultivated lands included *dalani mati*[16] or low lying marshy lands and *jangalajami*[17] or jungle lands; but these were also subsequently brought under cultivation. Most of the *baotalis* (lands for *bao* rice cultivation), *ahutalis* (lands for *ahu* rice cultivation) and *sariyahtalis* (lands for mustard cultivation) were lands thus cleared from jungle and marshes.

Lands were also classified as per nature of its soil and products. Epigraphs mention *minnik*,[18] *lalmati*,[19] *nilmati*, [20] *changana bhumi*,[21] *ajajangla*,[22] *hachilajangala*,[23] *bihijami*,[24] *svamillik rupit*,[25] etc. The meaning of all these terms cannot be known. While *lalmati* might have meant red soil, *nilmati* possibly meant lands for indigo cultivation, whereas *ajajangala* and *hachilajangala* were different types of jungle lands.

Bari and *vasti* constituted in the first part of the Ahom rule a class of non-revenue paying lands; but undoubtedly they formed an important class of lands. The epigraphs of the time mention in a number of occasions gift of *bari* lands to the Brahmanas.[26] These were high lands and were suitable for gardening and dry crop cultivation like brinjal, and nursery bed for growing seedlings. Hence some part of the *bari* lands were used for *kalani* (or plaintain garden), *kathiyatali*[27] (or nursery bed for paddy), *bengena tali* (or brinjal garden) and *sak* or *sakani bari* (vegetable garden). *Bari* lands were known as *arajami* or *arakata mati* (ancestral land) meaning legally inherited lands to which the occupant had his full ownership and nobody, even the king, could reclaim it.[28]

It is to be noted that the Ahom rulers considered every sort of land, including marshy lands, forests, *bil* or ponds and unused waste lands as useful. It is testified by the epigraphs, which while recording gifts of lands mentioned their nature and classified them in detail like cultivated, cultivable, jungle, marshy, dry or high, *da* or low-lying plots and included in the list even the ponds.[29] The *buranjis* mention that the rulers gave some of the *bils* in the border areas in lease to the hill tribes.[30] It is, therefore, clear that everything in nature—lands, forests, mines, ponds and rivers—all were taken by the Ahom kings as revenue paying sources and, therefore, they took special care to make them productive. Hence it is in this context that large scale division of lands in the medieval period can be explained.

Land Grants

Like the rulers of ancient Assam,[31] the Ahom kings also used to gift lands to the Brahmanas. Later they made land endowments also to different religious establishments like temples, *Satras, dargahs* etc. The records show that gifted lands included, besides cultivated *rupit* lands, lands of every other category mentioned above. The feature of the donated lands was that when gifted they became revenue free.[32] They were variously called *dharmottara, devottara, brahmottara nankar* and *pirpal* depending upon the nature of establishment to which they were made. Gait has mentioned as many as forty eight copper plate grants gifting lands to the Brahmanas and religious institutions by the Ahom kings which date from the reign of Gadadhar Singha.[33] It may be possible that the rulers even before Gadadhar Singha issued such land grant epigraphs. For, an Assamese chronicle called *Asam Buranji Sar* mentions that king Pratap Singha introduced the system of making land grants to Brahmanas and religious institutions under different categories such as *brahmottara, devottara* and *dharmottara.*[34]

The Ahom kings made large gifts of lands to the *Satras* or Vaishnava monasteries records of which are found beginning with the reign of king Jayadhvaj Singha and continued till the reign of Chandrakanta Singha (1811-1822). King Jayadhvaj Singha established the Auniati Satra and endowed it with extensive lands.[35] He made similar gifts to the Dakshinpat Satra also. Following him, his successor Chakradhvaj Singha (1663-1669) gifted eight hundred *puras* of revenue free lands to the Dakshinpat Satra on the occasion of his accession to the throne.[36] The later Ahom kings like Siva Singha (1714-1744) and Pramatta Singha (1744-1751) and their successors also donated large areas of lands.[37] The rulers of Assam gifted lands also to the Muslim *pirs* and their establishments. It is mentioned in a chronicle that Gadadhar Singha donated lands to one Muslim *pir* known as Azan Faqir in the mouth of the Dikhow river in the year 1684.[38] The *Padshah Buranji* (*Annals of the Delhi Badshahate*) mentions many land grants made to Muslim religious heads and their establishments in Assam. It mentions that in the year 1780, king Lakshmi Singha being pleased in his conversation with a Muslim Faqir called Anwar Hazi, granted him revenue free lands along with a number of servitors.[39] The same king granted lands to a number of Muslim religious establishments called *makam* or *dargah* in Kamrup.[40] In a like way when Kamrup was under the rule of the Mughals, the Mughal rulers also gifted lands to both Hindu and Muslim religious establishments. For example, emperor Auranzeb gifted lands in September, 1667, to the two priests of the Umananda temple at Pandu in Guwahati within the Dhakhinkul Sarkar.[41] The emperor also confirmed a grant of *madad-e-maas* or *madammas* of 2 *bisas* of land, i.e., 26 *bighas* to two priests of the Kamakhya temple which was originally gifted by Allah-Yar-Khan and Hasan Kandahari Khan, the two *Faujdars* of Guwahati.[42]

The Ahom rulers were patrons of learning and encouraged scholars to create literature by giving lands and all kinds of facilities. Thus the forefather of Ruchinath Kandali who translated the *Markandeya Chandi* into Assamese verse during the reign of Rajeswar Singha (1751-1769), was established by Rudra Singha (1696-1714) in his new capital city at Rangpur with grant of lands and other amenities.[43] This king gifted large plots of land to one Kaivalyanandan, the *Satradhikara* of the Dihing Satra, and patronised Sridharadeva of Telpani Satra for his proficiency in music.[44] The encouragement of music and learning by the Ahom kings led to the invitation of scholars from outside and their establishment within the kingdom with grants of land and servitors.[45]

The gifted lands contained certain characteristics, all or some of which are recorded in the copper plates. The most important of these were that the donor king almost permanently alienated the lands to the donee so that he and his descendants could enjoy the land for generations. It is found that even when the plates were lost, the grant was renewed by the succeeding rulers.[46] In one of his copper plate epigraphs recording grant of lands, king Siva Singha mentions that he would never exercise his power of collecting revenue and other taxes from the area and would refrain from even punishing the thieves there.[47] The donee would also obtain the property of those living within the donated area who died without leaving any heir to inherit them.[48] The grants further mention that no individual would be allowed to interfere in the donee's right and jurisdiction over the donated land. If anyone tried to do otherwise he was severely punished.[49] This, at the same time, implies that in spite of all rights being relinquished, the king retained his overall supervision over the gifted lands and the right of adjudication. A number of epigraphs mention that although the king gave up all rights over the gifted land, he retained his right to punish (*rajdanda*) the offenders violating the regulations of the gifted lands.[50] It may, also be noted that although the donee was given ownership he had no right to sell the land.

Beneficiaries of the land grants were mostly the Brahmanas who did not plough the land by themselves. Considering this aspect, kings also gifted labourers along with the lands. These labourers or *paiks* were then exempted from giving regular manual service to the state. It is recorded that king Rudra Singha, by two of his grants gifted along with lands 80 *paiks* as servitors to the Umananda temple.[51] King Siva Singha once gifted 3 *paiks* to the Siva temple at Dergaon, and on another occasion 17 families of *paiks* to the Barpeta Satra.[52] Rajeswar Singha by a grant of the year 1637 *Saka* (1765 A.D.) gifted 5¼ *gots* or 21/16 of *paiks* to a Brahmana called Kaviratna Bhagawati.[54] This king made gifts of large amounts of lands and servitor-*paiks* to different Brahmanas and temple institutions.[55] The donated *paiks* or their families had their names deleted from the regular list of the *paiks* who were to render service to the state.[56] This indicates that those *paiks* had no obligation towards the state and after being donated they became completely attached to the donee or the temple or the *Satra* as servitors. It can be presumed that grant of large number of *paiks* ultimately drained the state treasury in form of loss of man-power-revenue. The fact that Rajeswar Singha reduced the strength of a *got* of *paiks* from four to three,[57] can be explained in this light. Moreover, as a result of the gifts, a considerable part of the state lands went to the hands of the donees. Table 2.1 would show only a fraction of the lands thus alienated during the time of Siva Singha, Pramatta Singha and his successor Rajeswar Singha.

Table 2.1 shows that the gifted lands whether individual or institutional (religious) formed sufficiently big holdings in comparison with the *paik's* share of two *puras* of land which too could be reclaimed by the state after the death of the *paik* or in his inability to cultivate it.[58] Records show that the family of Krishnaram Bhattacharyya who was established by king Rudra Singha and was put in charge of the Kamakhya temple at Nilachal in Guwahati, was said to have possessed as free gift about half of the total cultivated and cultivable land area of the entire

TABLE 2.1: Some Largest Land-grant Records in the 18th Century, Assam

Name of the Donor & the Date of Gift	*Name of the donee/ Institution*	*Area of Land*	*No. of Paiks Granted*	*Nature of the Grant*
Siva Singha 1637 *Saka*	Brahmanas of Kamakhya, Pandunath etc.	400 *puras*	—	*Brahmottara*
Siva Singha 1639 *Saka*	Six Brahmanas	80½ *puras*	—	- ditto -
Pramatta Singha 1669 *Saka*	Ramananda Vidyalankar of Parbatiya Goswami's family	4488 *puras*	—	- ditto -
Rajeswar Singha 1676 *Saka*	Dipteswar temple	490 *puras*	many	*Devottara*
Rajeswar Singha 1679 *Saka*	Dirgheswari temple	426 *puras*	53½ *got*	- ditto -
Rajeswar Singha 1686 *Saka*	Dhananjoy Sharma (*Rajvaidya*)	120 *puras*	—	*Brahmottara*
Rajeswar Singha 1686 *Saka*	One Mahendra Brahmana (a *visharad*)	210 *puras*	—	- ditto -

Source: *PSN*, Ins. Nos. 16 and 17; *Banhi* (a monthly Assamese journal), Vols. XV-XVI; E.A. Gait, *Report on the Progress of Historical Research in Assam*, Shillong, 1897; *MEAG*, pp. 54f; Rajkumar, *op. cit.*, p. 439; Bhuyan, *Rajeswar Singha*, pp. 275f.

Kamrup region.[59] While preparing the *Gazetteer of Eastern Bengal and Assam*—the later including the states of the present north-east India in the beginning of the 20th century, B.C. Allen records that till that time, the Dakshinpat Satra possessed 12,000 acres of revenue-free land received earlier as gift from the Ahom rulers of Assam.[60] The same record mentions the possession of the Auniati Satra of 22,000 acres and that of Garamur Satra 40,000 acres.[61] Allen records that out of a total settled area of 25,62,000 acres, of Assam proper, 1,89,000 were half revenue paying land and 81,000 acres were revenue free.[62] One important aspect of the gifted lands was that a major part of them was cultivable *gamati* or *rupit*, i.e., wet rice lands.

Land Assignments

Besides religious and cheritable purposes, lands were also given to the state officers as assignments in lieu of pay. Beginning from the top cabinet ministers—the Buragohain, the Bargohain, the Barpatragohain, the Barbaruah and the Barphukan, a host of officers of both higher and lower ranks including royal favourites and retainers used to receive from the king for their maintenance land assignments and labour allotments.[63]

The three great *Gohains* who were holding power next to the king, were in fact, rulers of certain territories of the kingdom known as *Bilat* or *Rajya* or *Des*.[64] The Buragohain ruled over the territory extending from Sadiya in the east to the river Gerelua in the west on the north bank of the Brahmaputra.[65] While the Bargohain's territory lay to the west and extended upto the river Burai,[66] that of the Barpatragohain lay in between the *Bilats* of the Bargohain and the Buragohain.[67] Each of the *Gohains* was supposed to command 10,000 *paiks*[68] from their respective territories. But the *Gohain* who was considered as the seniormost was given 1,000 *paiks* more in addition to his normal allotment.[69] David Scott computed that the value of these allotted *paiks* in terms of money would be around Rs. 90,000 per annum.[70] Besides commanding their *paiks* in the battle-field, the *Gohains* could use these men in clearing waste land, which in course of time, they converted into cultivable *rupit* class then known as *khats*, i.e., farms. The *khats* were of considerable sizes ranging usually from 200 to 900 *bighas* of *rupit* land.[71] The *khats* or farms were considered as their personal property, rent-free and inheritable.[72] A *Gohain* resided usually in the capital till he was holding his office and in that capacity possessed, in addition to the *khats*, certain amount of homestead, orchard and cultivable lands free of rent. But he was to relinquish the same after his tenure was over.[73] During the period of their service a *Gohain* or any officer could utilise the entire manual revenue to which he was entitled for his service to the king. Besides, he was receiving as gifts some items in kind from the *paiks* under his control, a small amount of which was deposited to the royal treasury.[74]

The significant point regarding the great *Gohains* was that except for supplying *paik* soldiers to the king in times of war, they were independent of the king almost in every aspect of administration.[75] The other important aspect was that the *khats* or revenue-free assignments acquired through allotted *paik* labourers could be retained by them even after the expiry of the term of their office or by their descendants after their death. The third significant aspect was that the *paiks* or the *likchous* (personal attendants) discharged all necessary services to supply the *Gohains* with their daily necessaries so that their money expenditure was very minimum.[76]

As we have stated earlier, there were two other important state officers of cabinet rank just below the three *Gohains* known as the Barbaruah, and the Barphukan. The Barbaruah was the chief executive officer of the state exercising his power from Sadiya to Kaliabar in upper Assam and the Barphukan was the king's deputy who ruled over the lower Assam region—Kaliabar downwards, with his headquarters first at Kaliabar and later on at Guwahati.[77] The Barbaruah and the Barphukan commanded 14,000 *paiks* each on the king's behalf and each of them received in lieu of pay 7 per cent of this number, i.e., 980 *paiks* for their personal use.[78] Those *paiks* cultivated and worked in the *khats* for their masters. Besides, they could also collect fines for offences committed by the *paiks* in their jurisdiction, and contributions from junior officers on the occasion of their being appointed or promoted to a *pad*, i.e., a new post.[79] In addition to these prerogatives and land assignments, the Barphukan received 1,000 *puras* of rent-free lands from the king.[80] According to S.L. Barua, "such prerogatives together with wide and autonomous power" sometimes "induced a Barphukan to attempt to acquire suzerain power".[81]

Among the recipients of assignments were the royal princes appointed as governors to rule over certain territories of the kingdom known as *mel*. These governors were therefore called *Meldagia* or *Melkhowa Raja*.[82] The governors of the three territories of Charing, Tipam and Namrup were, called *Raja* and were considered more important than other *Melkhowa Rajas* or princes in the sense that the heir to the throne was selected from among them, the Charing *Raja* being usually made the heir apparent or *Yuvaraj*.[83] The *Melkhowa Rajas* used to appoint smaller officers in their territories and collected revenues and other taxes—which they themselves appropriated.[84] There were two *Melkhowa Gohains* or princes junior in rank to the *Rajas* known as *Majumelia Gohain* and *Sarumelia Gohain* who were also assigned areas of land for their own appropriation.[85]

The other significant group of assignees were the royal ladies—the king's consorts, the queen-mother, the king's grand-mother, step-mothers, sisters, daughters, nurses (*dhai*) and daughters-in-law. These ladies received from the king "sufficient means for their comfortable maintenance" including large plots of land as grants.[86] Such grants were managed by some officers placed under a *Phukan* or a *Barua*.[87] The biggest and significant estate or *mel* was that of the chief queen known as the *Raidangia mel*.[88] Other ladies received *mels* of different extents. Maniram Dewan has noted that the number of such *mels* increased to twelve during the reign of king Rajeswar Singha.[89] Unlike the *khats* of the princes and the ministers, the *mels* could not be enjoyed on hereditary basis.[90] It should be mentioned here that the systems of granting lands to the palace ladies were there in the kingdom of Pong in Upper Burma wherefrom the Ahoms left for Assam.[91] It is therefore presumable that this system of land assignment to the royal ladies was inherited by the Ahom rules from their original home-land. It may be mentioned here that the system of assigning lands to the royal ladies known as *inam* was also prevalent among the Mughals. The grantee here could enjoy the lands without holding any post.[92] For example, princess Jahanara, the daughter of emperor Shahjahan (1627-1658), was granted by her father such *inam* in Surat.[93] But granting of *inam* in large scale to almost every royal lady like that prevalent in Assam, was not found among the Mughals.[94]

There were three governors known as the *Datiyalia Gohains*: they were *Sadiyakhowa Gohain*, the *Marangikhowa Gohain* and the *Salal Gohain* ruling over the frontier provinces of Sadiya, Marangi and Sala respectively.[95] The governors were recruited from among the family members of the three great *Gohains.*[96] Like other officers and members of the royal family, they also received from the king certain prerogatives including land and *paiks*. There is, however, no specific records as to the number of *paiks* allotted to them. It appears that these governors also received land and *paiks* in the same way as the other *Melkhowa Gohains*,[97] and were to command their *paik* soldiers in times of war on behalf of the king.

At the lowest of the rank of the *Gohains* were two other *Gohains* known as *Kajalimukhia Gohain* and *Jagial Gohain*. The latter served under the Barbarua and looked after the state's relations with the seven bordering tribal chiefs known as *Sat Raja* at Jagi in Nagaon.[98] The *Kajalimukhia Gohain* served under the Barphukan and maintained relations with Jayantiya and Dimarua.[99] Each of them commanded one thousand *paiks* and obtained land and other facilities.[100]

Below these governors and officers there were hundreds of other civil and military officers known as *Phukan, Barua, Rajkhowa, Hazarika, Saikia* and *Bara*.[101] While a *Phukan* was in charge of 6000 *paiks*, a *Rajkhowa* or a *Barua*[102] commanded 3000 *paiks*, a *Hazarika* 1000, a *Saikia* 100 and a *Bara* 20 *paiks*.[103] Each of them received as assignment 10 per cent of the total number of *paiks* allotted to them for command.[104] Francis Hamilton writes that all "these officers are allowed lands free of rents which are cultivated by that proportion of the *paiks* that is allowed to work on their account, and each receives presents from the men and officers that are subordinate to his authority".[105] P. Naobaicha Phukan has put the amount of lands allotted to the higher *Phukans* and the

Rajkhowas as 700 and 500 *puras* each.[106] The rest, excluding all prerogatives and assignments, belonged to the king. He "employs a vast number of men to work in his farms, gardens, fisheries, mines, arsenals and manufactories and to man and construct his vessels, who all labour without any expense to the treasury".[107]

Land Ownership

The Ahom state possessed some peculiar characteristics so far the land ownership was concerned. From what is recorded in the *buranjis*, it appears that the great *Gohains* and the members of the seven blue-blooded Ahom families called the *Satgharia* Ahoms[108] could claim ownership of the state land by dint of their ancestral participation with the king in founding the kingdom, and therefore they treated the state-lands as common property of them all. An interesting reference to the claims made by one of the members of the original Ahom families who accompanied Sukapha and helped him to establish the kingdom here, can be found in the *Satsari Asam Buranji*. It is mentioned in this chronicle that one Hemena who belonged to such a noble family and was a resident of Gargaon, the Ahom metropoly, did not permit the king Suklenmung *alias* Gargaya Raja to reclaim his plot while requisitioning lands for the construction of a rampart (*garh*) around the palace.[109] The Ahom regulation of distribution of lands among the royal dignitaries contain that the Buragohain and the Bargohain possessed enormous lands not merely by dint of the royal favour, but because of their ancestral participation (*arakata*) in the process of foundation of the state.[110] Such rights were inalienable and permanent.

For common people, there were two types of ownership—permanent and temporary. Lands of permanent ownership consisted of the homestead and its surrounding garden area known as *bari*.[111] While the chronicles do not contain any reference to the permanent right of the peasantry over the homestead, early British records testify to this fact. For example, David Scott, Jenkins and Mills—all of them found that in Assam immediately after the overthrow of the Ahom monarchy every individual family was possessing their paternal homestead land on permanent basis.[112] It is also mentioned that this plot of land could not be reclaimed by the state, i.e., the king who was in all practical purposes the owner of the soil.[113] In this connection A. Guha rightly points out that this policy of giving permanent right to the peasants over his homestead and garden was inherited by the Ahom rulers from their ancestral homeland in South-east Asia. He points out to an inscription of a 13th century Tai ruler called Ramkhamhaeng (1292 A.D.) which among others, mentions that the homestead is inheritable to the discendants.[114] It runs as follows:

> "If a common man, a noble or a chief fell sick or died, the home of his ancestors, his clothings, his elephants, his families, his rice granaries, his slaves, the areca palm plantations of his ancestors were all transmitted to his children".[115]

This right over homestead lands was, however, a "right of hereditary possession only and not of alienation of their holdings".[116]

During the earlier periods, the peasants were not to pay any revenue for their homestead plots, but in later part of the Ahom rule, peasants were made to pay Re. 1 as rent per *pura* of land per annum,[117] which was, in fact, a half of the usual rate per *pura* of *rupit* land. It appears that the policy of assigning permanent right over the homestead plot and to make it almost revenue-free, was aimed to make the peasants live in peace and undisturbed so that they could be smoothly engaged in the state works without any hindrance. That was in fact, a way of encouragement to the peasants to engage themselves in production system. It is, however, interesting to note that the peasants could not sell his homestead even though it was his personal property. There is an inscription now preserved in the Assam State Museum at Guwahati which records the sale and purchase of two plots of lands by some persons in present Sibsagar district at the rate of Rs. 4 per *pura*.[118] Although this inscription is not dated, it can be presumed that it was of a later date when the Ahom administration to limit the possession of individual plots at two *puras* of revenue-free land was no longer there to prevent any extension causing drainage to the state treasury. At the same time by the time of the sale, money was almost fairly circulated in transactions. Thus the events of land sale and purchase in Assam by making payment of money may be assigned to the close of our period only. The *Tungkhungia Buranji* also records an event of land sale in upper Assam in which a commoner sold a piece of land who was executed for this act during the reign of king Gaurinath Singha (1780-1795).[119] In all, however, the cases of land sale in upper Assam were very few. Excepting the above two cases, available contemporary records do not throw much light in this regard.

Land sale, particularly high lands or homestead, was not, however, uncommon in western Assam. There are a number of records testifying to this fact. The *sanad* issued by emperor Auranzeb in the year 1667 A.D. mentioning the gift of a piece of land was valued at Rs. 30.[120] An inscription of the year *Saka* 1608 (1686 A.D.) records the purchase of a plot of land by a higher state officer at Hajo in lower Assam.[121] Another copper plate inscription of the year *Saka* 1645 (1723 A.D.) also records the event of sale and purchase of a piece of 6½ *puras* of land in Kamrup[122] which was suitable for both orchard and cultivation of rice, implying a high or homestead land; its value being Rs. 25.[123] Besides epigraphic records of the period the *Katha Guru Charit* also gives a number of instances in which homestead lands were sold.[124]

Land sale in lower Assam may be explained in the light of two facts: (a) the Ahoms retained the entire system of the Mughal administration in the region in which land sale was not uncommon,[125] (b) the absence of the *khel* system[126] unlike upper Assam region in which lands were allotted to the *khels*, not to the individuals for distribution among the *paiks*. The collection of revenue in terms of money rather than in terms of labour, created a money market there and this had naturally encouraged sale of anything including lands in which a peasant had his authority.

Except the homestead plot which could be owned through inheritance, no land of whatever quality, belonged to the peasants individually. From what is found in the land grants of the period the king possessed the sole ownership of every kind of land—cultivated, cultivable and forest lands including ponds, rivers and mines. Extent grants mention how the rulers gifted these along with cultivated *rupit* lands to the Brahmanas or to the religious institutions.[127] The policy of expansion of agriculture in the state, however, put the peasants at liberty to clear jungle lands and bring them under cultivation as much as they could and retain them till they were not required by the state. But, then, when such lands became first class, i.e., *rupit*, the king reclaimed and distributed them among the *paiks* who were newly registered. In fact, a *paik* lost his right over the plot of land which he acquired by clearing the uncultivated waste as soon as it became cultivable. This, and the fact that the *paiks* were to regularly work for the state for about 3-4 months in a year, should have worked as inhibiting forces among the common peasants to open vast areas of new lands. But the volume of production of seasonal crops like mustard and pulses from the newly opened lands, as will be shown in a subsequent chapter, shows that inverse was the case in actual practice.

The next problem regarding peasants ownership of cultivable lands was that the state allotment of lands was conditioned by two factors; firstly, the land was given to the *khel* or community, not to the individual by name, to which he belonged; and secondly the *paik* peasant could enjoy the land so long as he was capable of paying revenues to the state in form of his service. This was how when a *paik* died or became in any way incapable of serving the state the king reclaimed his plot[128] and it was given to another *paik* through the *khel* to which he was newly registered. Thus a peasant was more a contract worker to the state than a free peasant with his own landed property. So it can be safely concluded that a peasant having neither the right to possess cultivable land permanently, nor any right to sale lands including his homestead (in case of upper Assam), nor having the right to leave the field to be inherited by his children unless they would become *paiks*, had in fact, no ownership of lands. Thus ownership of landed property ultimately rested with the state, i.e., the king, unlike in other parts of India where it varied in different circumstances.[129]

While peasants did not possess ownership rights over lands, the nobility including the state dignitaries, Brahmana donees and the heads of the *Satras* (*Satradhikaras*), the priests, and the *pandits* could possess large plots of land of every kind with proprietary rights over them. The Assamese chronicle already referred to, which mentions of one Hemena who did not concede to the demands of the king himself in leaving his plot of land for the construction of the rampart around the capital complex, reveals that this gentleman discarded the king's demand outright because the plot was owned by him by right of conquest. This shows that the members of the nobility by right of the joint conquest[130] possessed permanently some revenue-free lands and had proprietary rights over them. The other class of people who possessed such right were the donees. Lands gifted to the Brahmanas and the priests, the *Satradhikaras*, the Muslim *Pirs*, and the religious institutions were given by the king almost permanently. The grants speak that the donees and their descendants could enjoy the lands for many generations without limit,[131] i.e., these lands became the property of the donee, an inheritable by his succeeding generations. It so happened that gifted lands formed sometimes such autonomous territories inside the state that even the laws of the king governing his subjects had no effect in internal administration of such territories.[132] Thus the nobles and the other dignitaries as mentioned above which formed about 1 per cent of the total population[133] were not only possessed of lands, but also the ownership over them. The paradox is that none of these class personally tilled the lands or engaged themselves in agricultural works. On the other hand, they collected by way of gift and assignment a large number of workers to work in their field. It is mentioned earlier how thousands of *paiks* including *lickous* and slaves were attached to the households of the nobles and dignitaries depending upon their respective ranks. We shall discuss the fate of such bonded or gifted working class peasants in a subsequent chapter.

Ownership in the Tribal Areas

Besides the heartland of the Assam state, there were in its periphery a number of tribes who were till then not acquainted with developed ways of agriculture of transplanted *sali* variety. They were all adapted to jhuming and shifting cultivation. We have earlier mentioned, that the Marans and the Barahis, the first two aboriginal plain's tribes of the extreme south-eastern region in the valley to be subdued by the founder Ahom king Sukapha, presented them with the best of their articles, which consisted mainly of forest products and jhum crops.[134] The early British writers noticed that till the British occupation of Assam these tribes were used mostly to cultivation of *ahu* rice, which was so less productive and devoid of any surplus that it could not form an item of their presents. It should be mentioned that even in subsequent times the Ahom impact of developed agriculture was limited on these tribes which can be assumed from the fact that the Maran tribes used to supply the Ahom kings through the centuries of their rule, only with honey, wood, elephants and vegetable products.[135] The neo-Vaishnavite movement which spread among these tribes in the later part of the 17th century had some impact on their economy and culture, but could not bring a radical change in the method of their production. The early British records show that till at least the beginning of the 19th century tribal way of living predominated among them. Hannay in 1839 observed about the Marans as follows: "their lands are high, and their cultivation is 'ahu' crop of rice, once a year, and large crops of cotton and sugarcane but on account of scantiness of the population, compared to the extent of land capable of cultivation, their villages are scattered and the inhabitants are constantly emigrating to new sites, for the sake of richer and newer lands. . . ."[136] The Britishers thus finding large plots of land in the Matak territory lying open, reclaimed these for opening the tea gardens.[137] This shows that the methods of slash and burn cultivation and fallowing of lands for long period were still then practised by the tribes and they could retain blocks of lands for their communal use besides receiving *rupit* lands as *paiks* from the king. All these lead us to conclude that the Marans and the Barahis were having communal ownership of lands during our period of survey. The Ahom suzerainty over them could not do away with their tribal socio-economic structure.

Besides the Marans and the Barahis, there were among others, tribes like the Bodos or Kacharis, the Meches and the Garos whose population spread over almost the entire Assam region. While the main branch of the Kacharis formed a state in south-western part of the Brahmaputra valley contemporaneously with the Ahoms,[138] a number of the Kacharis, till the last part of the Ahom rule, were under the latter's suzerainty. The Garos were mostly governed by their own chiefs who were autonomous in their administration, but were under the Ahom sovereign power. Some of such principalities (*Rajya* or *Desh*) were those of Rani, Luki, Beltala, Naduar, Gobha, Dimarua, Bhagaduar, Mairapur etc.[139] Although no specific record to the system of their land and agricultural patterns are found from the early chronicles, it can be safely presumed that almost all of them were adapted to shifting cultivation with communal ownership of land.

In the neighbouring hill areas the tribes like the Nagas, Garos, Daflas (Nishis), Akas (Hrusso), Mishimis, Mikirs (Karbis) also practised shifting cultivation with communal ownership of land.[140] For example, in the Garo hills under the chief called *Nokna* there was the system of clan ownership of land called *A'khing*.[141] Recent survey of land systems in the Garo hills reveals that although the *Nokna* was the chief, land belonged to the clan; the chief was only the authority of allotting lands to the clan members for cultivation.[142] Being situated in the hills, the *A'khing* lands are generally utilised for jhum cultivation and some valuable crops are grown therein. For example, the short staple cotton of the Garo hills is well known. Other jhum crops included ginger, hill rice, maize and millets etc.[143] In Mizoram also chiefdom prevailed wherein all lands in a village for all practical purposes belonged to the community headed by a chief. It was the chief who distributed the lands among the village *Ramhuals* (i.e., cultivators) and collected a share of their produce as taxes.[144] In fact, the clan members did not own lands except over his piece of homestead which was inheritable.[145]

The Mikirs who were a hill tribe and lived in the areas covered by the present districts of Golaghat, Nagaon and North Cachar[146] within Assam, are accustomed to shifting cultivation with the use of fire, hoe and also plough.[147] In the period under discussion they were ruled by their chiefs under the suzerainty of the Ahoms, and even till date most of them shift from place to place in search of fertile cultivable land. They had not only communal ownership of land but even the institution of having a common granary. In the Naga dominated territories (now forming the state of Nagaland) land in most cases was owned by the community or by the clan under the supervision of the clan chiefs.[148]

Thus it can be said from the above that the king in the plains of Assam and the chiefs among the tribes of both the hills and the plains, had actual ownership of land. We have seen that for all practical purposes they had the upper hand in selection, retention, distribution and reclamation of lands. This gave them ample opportunity to collect land taxes in form of personal labour or produce of the soil and also to produce for themselves enough grains in their choosen and better quality lands.[149] While in the hills the chief allotted lands to the clan members only, in the plains of Assam, the ruler assigned lands to the state officers in lieu of pay and granted revenue-free lands to religious heads and priests and made provision for allotment of quality lands to the peasants. This measure helped in the development and strengthening of the state machinery on one hand and Hinduisation and spread of Brahmanical influence over the people including the ruling tribes on the other. This was how while the hill people retained tribal way of living, in the plains, tribal culture gradually eroded and feudal structure began to grow.

REFERENCES

1. Thomas Fitzsimmons, *op. cit*, also see *Supra*, p. 27, pp. 59f; G.A. Grierson, *Linguistic Survey of India*, Vol. 2, Calcutta, 1904, pp. 127ff; Powell, *op. cit*, p. 150; *MEAG*, p. 63.
2. *Mung-dun-Sun-Kham* means 'country full garden gold' and *Mung-ri-Mung-ram* means 'country deserted, country uninhabited'; see *HAG*, pp. 74, 78.
3. *DAB*, p. 1; *ABGB*, p. 10.
4. *AB*, p. 10.
5. *MEAG*, p. 63.
6. *Infra*, p. 144.
7. P.C. Choudhury, *The History of the Civilization of the People of Assam to the Twelfth Century, A.D.*, 2nd edn, Guwahati, 1966, p. 277; B.K. Barua, *A Cultural History of Assam*, Vol. I, 2nd edn, Guwahati, 1969, pp. 84f.
8. See for the system of requisition of personal service from the peasants Chapter IV of this work.
9. Among other epigraphs of the period mention of the term *rupit* is found in the *Copper Plate (CP) Land Grant Inscription of Konwar-bhag and Pubpar Pargana, Saka* 1677 (1755 A.D.); *C P Land Grant Inscription of Satsangi Satra, Saka* 1686 (1764 A.D.); *Saka* 1686 (1764 A.D.); *C P Land Grant Inscription of Bhubaneswari Temple, Saka* 17. . . . etc. See M. Neog (comp. and ed.), *Prachya Sasanawali*, Guwahati, 1974 (henceforth *PSN*), Ins. Nos. 64, 73, 91. The term *rupit* is derived from the Sanskrti word *ropana*, i.e., to plant, to root up and then to transplant, because the rice is often sown in the nursery beds and the

seedlings are transplanted when these are called *Sali*. In Assamese there is a word *rowa* which also means to transplant. In that sense the terms *rupit, ropana, rowa* and *rowati* have the same roots in their origin. Also see Boden Powell, *The Land Systems of British India*, Vol. III, Indian Print, Delhi, 1974, p. 416fn.

10. William Robinson, *A Descriptive Account of Assam*, Reprint, Delhi, p. 201 (henceforth *DAR*); *ADB*, p. 496.
11. *CP Land Grant Inscription of Sundaridiya Satra, Saka*, 1660 (1738 A.D.) and the *CP Land Grant Inscription of Aswakranta Devalaya, Saka* 1661 (1739 A.D.). See *PSN*, Ins. Nos. 37, 41.
12. For the term see the *CP Inscription of Saktipara Village in Khata Pargana, Saka* 1677 (1755 A.D.) and the *CP Land Grant Inscription of the Mahajan of the Bajali Pargana, Saka* 1687 (1765 A.D.), *PSN*, Ins. Nos. 63, 76. Also see *KGC*, p. 74.
13. *DAR*, p. 205; *ADB*, p. 497; A.J.M. Mills *Report on the Province of Assam* 2nd edn, Guwahati, 1984, p. 2.
14. *CP Land Grant Inscription of the Math of Kshudra Makhibaha Village* (Kamrup), *Saka* 1692 (1770 A.D.); *CP Brahmottara Land Grant Inscription in the Bichankuchi Village* (Kamrup), *Saka* 1665 (1743 A.D.); *CP Brahmottara Land Grant Inscription in the Pachimpar etc. Pargana* (Kamrup), *Saka* 1710 (1788 A.D.). See *PSN*, Ins. Nos. 79, 136, 141.
15. *PSN*, Introduction, p. 120.
16. *CP Land Grant Inscription of Opar Barbhag etc. Pargana* (Kamrup), *Saka* 1706 (1784 A.D.); *ibid.*, Ins. Nos. 76 and 84.
17. *CP Inscription of Dharmottara for Satsangi Bhakats in Bangsar etc. Pargana, Saka* 1663 (1741 A.D.), *ibid.*, Ins. No. 133.
18. *CP Inscription of Land Grant of Asvakranta Temple, Saka* 1611 (1719 A.D.), *ibid.*, Ins. No. 41.
19. *CP Inscription of Kaviratna Saraswati, Saka* 1687 (1765 A.D.); *CP Inscription of Anowar Faqir, Saka* 1702 (1780 A.D.). See *ibid.*, Ins. No. 77, 83.
20. *CP Inscription of Opar Barbhag etc. Pargana, Saka* 1706 (1784 A.D.), *ibid.*, Ins. No. 85.
21. *Land Grant Inscription of Kamakhya, Pandunath and Ugratara Temple, Saka*, 1637 (1715 A.D.), *ibid.*, Ins. No. 16.
22. *Ibid.*, Ins. No. 133.
23. *Ibid.*, Ins. No. 136.
24. *Ibid.*, Ins. No. 133.
25. *CP Inscription of Brahmottara in the Nambarbhag Pargana, Saka* 1663 (1741 A.D.), *ibid.*, Ins. No. 132.
26. *CP Inscription of Dharmottara grant in Khata Pargana, Saka*, 1664 (1742 A.D.), *ibid.*, Ins. Nos. 4, 76; also see Ins. No. 77.
27. *Ibid.*, Ins. No. 136; also see Ins. No. 63.
28. *CP Land Grant Inscription of Sandheli Namghar, Saka* 1695 (1773 A.D.), *ibid.*, Ins. No. 81; *SAB*, p. 20.
29. *DRV*, v. 549; *Document of Dharmottara and Nankar Land Grant in Bangsar Pargana, Saka* 1743 (1821 A.D.); *CP Land Grant Inscription of Sundaridiya Satra, Saka* 1660 (1738 A.D.). See *PSN*, Ins. Nos. 102 and 37. For classes of lands gifted in other parts of India. See R.S. Sharma, *Bharater Samanta Tantra* (Bengali Translation of *Indian Feudalism*), Second edn, Calcutta, 1985, pp. 92ff.
30. *ABHB*, pp. 40f; *ABS*, p. 27.

31. See for detail M.M. Sharma's *Inscriptions of Ancient Assam*, Guwahati University, 1978; D. Nath, 'Aspects of Agrarian Relations as Reflected in the Copper-plate Grants of Ancient Assam', *JHR*, Vol. IV, Dibrugarh University, 1991, pp. 40ff.
32. All the land grant epigraphs of medieval Assam testify to the fact that when a piece of land was gifted it was made free from all sorts of regular and irregular taxes. See, for example, the land grant epigraphs collected in the *PSN* and those in the *Inscriptions of Ancient Assam*. In Mughal India such lands were known as *Madad-e-maas*. See Habib, *op. cit*, pp. 298f.
33. *HAG*, p. 184fn.
34. *ABS*, p. 28.
35. T. Sharma, *Auniati Satrar Buranji*, Auniati Satra, 1975, p. 12.
36. B. Sharma, *Dakshinpat Satra*, Guwahati, 1964, p. 42.
37. See *CP Dharmottara Land Grant Inscription to Patbausi Satra* during the reign of Siva Singha, *Saka* 1639 (1717 A.D.); *CP Land Grant Inscription to Dakshinpat Satra* during Pramatta Singha, *Saka* 1671 (1749 A.D.); *CP Inscription of Barpeta Satra, Saka* 1657 (1735 A.D.); *CP Land Grant Inscription to Patbausi Satra* during Lakshmi Singha, *Saka* 1694 (1772 A.D.); *Land and Servitors Grant Inscription to Dakshinpat Satradhikara* during Gaurinath Singha, *Saka* 1707 (1785 A.D.); *Document of Dharmottara Land Grant to Aibheti Na Satra* during Chandrakanta Singha, *Saka* 1744 (1822 A.D.), *PSN*, Ins. Nos. 18, 31, 52, 80, 86, 104.
38. Cf. M. Neog, *Asamiya Giti Sahitya in Maheswar Neog Rachanawali*, Vol. I, Dibrugarh, 1986, pp. 396ff.
39. S.K. Bhuyan (ed), *Annals of the Delhi Badshahate*, Guwahati, 1947, p. 15. (For the text of the epigraph see *PSN*, Ins. No. 83).
40. *Ibid.*
41. *Ibid.*, p. 18.
42. *Ibid.*, p. 17. S.K. Bhuyan's presumption of a *bisa* as being equal to a *bigha* is possibly not correct. H.N. Choudhury in his *The Coch Behar State and Its Land Revenue Settlement* (Coch Behar, 1903) mentions that it was equal to thirteen *bighas* of land. See p. 549.
43. Cf. Bhuyan, *Rajeswar Singha*, p. 199.
44. *Ibid.*, pp. 186f; D. Nath 'Asamat Sahitya Sristit Rajpristha-poshakata' in *Sahitya Sanskriti Saurabh*, (ed), D. Kakati, Sibsagar, 1993, p. 98.
45. See for detail *Sahitya Sanskriti Saurabh*, pp. 88ff.
46. See the *CP Inscription of Siddheswar Temple, Saka* 1645 (1723 A.D.), and the *CP Inscription of Opar Barbhag etc. Pargana, Saka* 1706 (1784 A.D.). See *PSN*, Ins. Nos. 24, 85.
47. *Ibid.*, Ins. No. 24.
48. *Ibid.*, Ins. Nos. 24, 31, 37.
49. *Ibid.*, Ins. Nos. 31, 37.
50. *Ibid.* For example, see Ins. Nos. 24, 30, 31, 37, 83, 85, 91. Also see Habib, *op. cit*, p. 235; R.S. Sharma, *op. cit*, pp. 2ff; *JHR*, Vol. IV, p. 52. The primary regulations governing those lands included (1) payment of all kinds of taxes and dues to the donee instead of to the state by the residents of the gifted areas without protest, (2) not to disturb the donee by any person or state officers in the enjoyment of the land gifted, and (3) to maintain loyalty to the state by the donee and his discendants and to pray for the good and religious merit of the king.
51. *The Land and Servitors Grant Inscription of the Umananda Temple, Saka* 1617 (1695 A.D.); *The Land and Servitors Grant Inscription of the Umananda Temple, Saka* 1619 (1697 A.D.). See *PSN*, Ins. Nos. 13, 14.

52. *The CP Inscription of the Siva Temple at Dergaon, Saka* 1656 (1734 A.D.); *PSN*, Ins. No. 30.
53. *Ibid.*, Ins. Nos. 31, 76.
54. *Ibid.*, Ins. No. 76.
55. See the Table 2.1.
56. *PSN*, Ins. No. 30; Also see *The Land and Servitors Grant Inscription of the Bengenaati Satra, Saka* 1699 (1773 A.D.) Ins. No. 82.
57. *HAG*, p. 239; *ADB*, p. 495; *Mill's Report*, p. 2; S. Rajkumar, *Itihase Sowara Chasata Bachar*, Jorhat, 1980, pp. 157f. According to Harakanta Barua Sadaramin the Moamaria revolt which caused much killing of peasantry in the state, led also to reduction of number of *paiks* in a *got*. See *ABHB*, p. 120. It may be possible that the gradual increase of the state officers and their *paik*—allotments also led to reduction of the *paiks* in each *got* in later times.
58. Infra, p. 69.
59. Bhuyan, *Rajeswar Singha*, p. 118. Also see Appendix-D.
60. B.C. Allen, E.A. Gait, *et. al Gazetteer of Bengal and North-East India* (Originally *Provincial Gazetteer of India—Eastern Bengal and Assam*), Reprint, Delhi, 1979, p. 585.
61. *Ibid.*, pp. 583, 586.
62. *Ibid.*, p. 103.
63. *AAH*, p. 24; A. Mackenzie, *The North-East Frontier of India*, Indian Reprint, Delhi, 1981, p. 6.
64. *ABSM*, p. 63; *ABGB*, p. 108; *ABHB*, p. 118, Rajkumar, *op. cit*, pp. 310ff; Jatin Bargohain, *Asamar Arthanatik Itihas*, Jorhat, 1985, p. 6.
65. Rajkumar, *op. cit*, p. 370.
66. *Ibid.*, p. 374.
67. *Ibid.*, p. 372. Earlier the Buragohain and the Bargohain received from king Suhungmung the territories around Dihing, and Sadiya as their respective *Rajyas*. *SAB*, p. 68.
68. *Note by David Scott* in Mackenzie, *op. cit*, p. 378; Bargohain, *op. cit*, pp. 7ff. Each thousand of *paiks* thus allotted formed a unit called *Hatimur*. See *SAB*, pp. 67f, *AAH*, p. 23.
69. *ABHB*, p. 95; Bargohain, *op. cit*, p. 8. Captain Welsh, however, mentions that the number of the additional *paiks* was 2000; see Mackenzie, *op. cit*, p. 379.
70. *Note by David Scott*, Mackenzie, *op. cit*, p. Appendix-378.
71. *Ibid.*
72. *Ibid.*, p. 6; U.N. Gohain, *Assam under the Ahoms*, Jorhat, 1942, p. 118.
73. Wade, *op. cit*, Introduction, p. XV.
74. *Report of Captain Welsh* in Mackenzie, *op. cit*, pp. 378f.
75. *Ibid.*, p. 378; Gohain, *op. cit*, p. 118.
76. *AAH*, p. 23.
77. Wade, *op. cit*, Introduction, p. XVII; *HAG*, p. 110.
78. *Note by David Scott*, Mackenzie, *op. cit*, p. 379; *HAG*, p. 236; *ADB*, p. 531; Gohain *op. cit*, pp. 101ff; Robinson and P. Naobaicha Phukan put the number of *paiks* allotted at 100 or 25 *gots* only. See *DAR*, p. 19; *Asam Buranji* (Ms).
79. *Note by David Scott* in Mackenzie, *op. cit*, p. 379.
80. *Asam Buranji* (Ms) by P. Naobaicha Phukan; L. Gogoi, *Tai Sanskritir Ruprekha*, 3rd edn, Calcutta, 1985, p. 188; *CHAB*, p. 390.
81. *CHAB*, p. 390.
82. *ADB*, p. 537; Bargohain, *op. cit*, p. 11; Rajkumar, *op. cit*, p. 175; Gohain, *op. cit*, p. 108.

83. *ADB*, p. 537; *CHAB*, p. 391.
84. Bargohain, *op. cit*, pp. 11f.
85. *Ibid.*, p. 123; *ADB*, p. 423.
86. *Jenkin's Report to C. Beadon*, Secy. to the Govt. of Bengal, Judicial Dept., dated 5th July, 1853; Cf. *Mills Report*, p. 638.
87. *Buranji Vivek Ratna, II*, (Ms); Bhuyan, *Rajeswar Singha*, p. 229.
88. Bhuyan, *Rajeswar Singha*, p. 229. Considering the importance and dignity of this queen the British Company Govt. granted a pension of Rs. 200 per month for the chief queen of Purandar Singha after his death. See *Jenkin's Report*, 5th July, 1853, Cf. *Mills Report*, p. 638.
89. *Buranji Vivek Ratna, II* (Ms).
90. *ADB*, p. 430.
91. G.T. Bayfield, *Narrative of a Journey from Ava to the Frontier of Assam*; Cf. U.N. Gohain, *op. cit*, p. 120; *Supra*, p. 23.
92. I. Habib, *Mughal Bharatar Krishi Vyavastha (1556-1707)*, being the translation into Bengali of *The Agrarian System of Mughal India (1556-1707)*, Calcutta, 1985, p. 274 (henceforth Habib (*B*).
93. Abdul Hamid Lahori, *Padshahnama*, Vol. II, p. 397; cf. *ibid.*
94. Even Habib who discusses this aspect almost in details had cited one example only. See *ibid.*
95. *Captain Welsh's Report* in Mackenzie, *op. cit*, p. 381; M. Martin, *The History, Antiquities, Topography and Statistics of Eastern India*, Vol. V, First Indian Print, 1976, pp. 623f; *ADB*, pp. 534f; Gohain, *op. cit*, p. 108.
96. *ABHB*, p. 119.
97. Martin, *op. cit*, p. 623.
98. Rajkumar, *op. cit*, p. 381, The *Sat Rajas* can not be identified for certain. The *buranjis* very often mention a group of *Panch Rajas* (five kings) along with the *Sat Rajas*. Both the groups included the Mech and the Garo chiefs of Kamrup, Nagaon and Darrang. See Martin, *op. cit*, pp. 619ff; Bhuyan, *Rajeswar Singha*, p. 240.
99. Martin, *op. cit*, p. 623; *ADB*, pp. 544f; Gohain, *op. cit*, pp. 108ff; Rajkumar, *op. cit*, p. 381.
100. *ABHB*, pp. 118f; Martin, *op. cit*, p. 623; Gohain, *op. cit*, pp. 108f. Also see the Appended part of the *Petition of Maniram Dewan* submitted to Mills on behalf of Prince Ghanakanta in *Mills Report*, pp. 610ff.
101. *ABHB*, pp. 118ff; *PAB*, p. 60; *ADB*, p. 419; Gohain, *op. cit*, p. 109.
102. A *Barua* was generally a civil officer, departmental or administrative head or deputy. But a *Rajkhowa* was ordinarily a governor of a territory wherefrom a levy of 3000 *paiks* could be recruited. See Bhuyan, *op. cit*, p. 10. Sometimes a *Barua* acted as head of departments in the absence of a *Phukan*, or as a second officer where the *Phukan* heads the department. See *Tungkhungia Buranji* (English) (ed). S.K. Bhuyan, 2nd print, Guwahati, 1968, Appendix III, p. 229 (henceforth *TB* (Eng.).
103. *ABGB*, pp. 207f; *PAB*, p. 60; Report of Captain Welsh in Mackenzie, *op. cit*, p. 380. According to Mills a *Bara* commanded 20 *gots*, not 20 individual *paiks*. See *Mills Report*, p. 2.
104. Wade, *op. cit*, Introduction, pp. xvf. The *KGC* mentions that a *Bara* received for his own use two *poas* (i.e., 2) of *paiks* out of the twenty he commanded which formed 10 per cent of that number (see pp. 465, 520). On the other hand, Francis Hamilton states that the officers were entitled to a commission of 5 per cent of

the *paiks* commanded by each of them (see *AAH*, p. 24). It is possible that in emergent situation like war when five hundred *paiks* were counted as forming a *hazar* or thousand; i.e., when a Hazarika, instead of commanding one thousand *paiks*, had commanded only five hundred, the percentage of *paik*-allotment was also reduced from earlier ten to five per cent. *ABSM*, p. 90; S.L. Barua, *The Last Days of the Ahom Monarchy*, Delhi, 1991, p. 263.

105. *AAH*, p. 23.
106. *Asam Buranji* (Ms).
107. *Ibid.*, p. 24.
108. *ADB*, pp. 504ff.
109. *SAB*, p. 20.
110. *Ibid.*
111. *DAR*, p. 203.
112. *Note by David Scott* in Mackenzie, *op. cit*, pp. 377f; also *Mills Report*, pp. 2f.
113. *Mills Report*, p. 2; W.E. Ward, *Assam Land Revenue Regulation*, 1886, Introduction, p. LIX; Mackenzie, *op. cit*, p. 6; J.N. Phukan while discussing the ownership of lands, remarks that in the traditional concept of Ahom kingship, the king was the owner of the kingdom, its soil and all other properties. See *The Economic History of Assam under the Ahoms* (Ms), Ph.D. Thesis, Guwahati University, 1973.
114. *MEAG*, p. 48.
115. Cf. *Ibid.*, pp. 43f.
116. Phukan, *op. cit.*
117. *Mills Report*, p. 2; Mackenzie, *op. cit*, p. 6.
118. See *PSN*, Ins. No. 156.
119. S.K. Bhuyan, (ed.), *Tungkhungia Buranji*, 2nd edn, Guwahati, 1964, p. 113 (henceforth *TB*).
120. See *Supra*, p. 47.
121. Preserved in the Record Room, Assam Secretariat, Guwahati, Ins. No. 50; T. Raychoudhuri and Irfan Habib (ed.), *The Cambridge Economic History of India*, Vol. I, Reprint, Delhi, 1984, p. (Appendix) 486.
122. Assam Records Room, Assam Secretariat, Ins. No. 28; *The Cambridge Economic History*, I, p. 486.
123. *Ibid.*
124. pp. 87, 42, 263.
125. *The Cambridge Economic History*, I, p. 247; S. Nurul Hasan, *Thoughts on Agrarian Relations in Mughal India*, New Delhi, 1973, p. 23.
126. Se *Infra*, p. 157.
127. See Table 2.1; *PSN* for detail; Nath, *op. cit*, pp. 45f.
128. Also see Phukan, *op. cit.*
129. See for detail R.S. Sharma, *op. cit*, pp. 112ff; *Habib(B)*, p. 120ff.
130. The top three families included those of the king, the Buragohain, and the Bargohain. Later on the family of the Barpatragohain also was added to the order. As per convention, the Ahom administration depended for its smooth running on the full co-operation of those families in which each one had its inherited share of voice. See *Report of Captain Welsh* in Mackenzie, *op. cit*, p. 378; *HAG*, pp. 235f; Gohain, *op. cit*, pp. 96ff.
131. For example, see the *Land and Servitors Grant Inscription of Bengenaati Satra, Saka* 1699 (1773 A.D.); *The CP Land Grant Inscription in Kachari Mahal etc., Saka* 1711 (1789 A.D.). See *PSN*, Ins. Nos. 82, 90.

132. *Supra*, pp. 48f.
133. See Chapter IV.
134. *DAB*, p. 101; *SAB*, pp. 6f; *ABSM*, p. 5; *ABS*, p. 10.
135. Groups of people specified for supplying different articles were known as *Hatichungi* (elephant supplier), *Maujogania* (honey supplier), *Kharijogania* (supplier of fire wood), *Rangjogania* (supplier of dying colour), *dhariboa* (mat maker) etc. See Bhuyan, *op. cit*, p. 207.
136. *Foreign Political Proceedings*, 14 August, 1839, Nos. 105; *MEAG*, p. 66.
137. *Political Consultation*, 14 August, 1839, No. 106; R.M. Lahiri, *The Annexation of Assam*, Reprint, Calcutta, 1975, pp. 202ff. See the Waste Land Rules in *Revenue and Judicial Letters*, 213-14, 14 March, 1837; H.K. Barpujari, *Assam in the Days of the Company*, Guwahati, 1963, p. 212.
138. *Supra*, pp. 16f.
139. See for detail, *AAH*, pp. 30ff; Martin, *op. cit*, pp. 619ff.
140. J.N. Das (ed), *Land System of North-Eastern India*, Vol. VI (Ms), Shillong.
141. *Ibid.*
142. *Ibid.*
143. *Ibid.*
144. N.E. Perry, *Luchai Customs and Ceremonies*; cf. Das, *op. cit*.
145. Das, *op. cit*.
146. Allen, *et. al*, *op. cit*, p. 154; E.T. Dalton, *Tribal History of Eastern India*, New Delhi, Reprint, 1970, pp. 53f; *DAR*, p. 311; Devi, *op. cit*, p. 145.
147. John Butler, *Travels in Assam*, second Reprint, Delhi, p. 139; *DAR*, p. 311; Devi, *op. cit*, p. 147.
148. B.B. Ghosh, 'Pattern of Use and Ownership of Land in Nagalands', in *Land Reforms and Peasant Movement*, (ed), A. Goswami, Guwahati, 1986, pp. 183ff.
149. In case of the Mizos, Perry mentions that the "Ramhuals select the slopes that they think will be the most suitable for jhuming for the year. The chief then gets the first choice and selects the land he wants for his own jhum. After him the *Ramhaul* select their jhum in order. . . ." cf. Das, *op. cit*. Earned in this way, Shihabuddin Talish records that the Ahom king Jayaddhvaj Singha also possessed one hundred and seventy granaries; cf. *HAG*, p. 134.

3

EXTENT AND METHOD OF CULTIVATION

Extent of Cultivation

Due to non-availability of sufficient contemporary records it is not possible to ascertain precisely the extent of cultivation in Assam in the period under this study. The problem naturally involves two aspects: the amount of land brought under cultivation and the volume produced. The Ahom kingdom at its zenith in the 2nd half of the 17th and throughout the 18th century covered almost the entire plains districts of present Assam excluding, of course, the Barak valley and the undivided Goalpara district of modern Assam.[1] J.P. Wade while writing on the geographical features of the state of Assam in the close of the 18th century speaks of its extent as being "bounded on the south-west by Bengal and Bisene (Bijni); on the north by the successive ranges of the mountains of Bootan, Aka, Duffla and Miree; on the south by the Garoo mountains, which rise to a greater height in proportion to their progress eastward, and change the name of Garoo, for that of Naga, above Coliabar (Kaliabar in Nagaon district now)".[2] The kingdom during this period was spread over about 500 miles in length and 60 miles in breadth. It thus covered an area of about 30,000 sq. miles.[3] Geographically the area as mentioned can be divided into three major regions: the north bank (*uttarkul*), the south bank (*dakshinkul*) and the island of the Brahmaputra called Majuli.[4] The entire belt is a region of alluvial land and exhibits a highly fertile soil throughout; even most of the hills surrounding Assam have some areas which are susceptible of cultivation. Also the *chapari* or the island belt which consists of the islands and the riverine tract of the Brahmaputra, being subject to rains and flood, were fertile enough to promote seasonal cultivation involving 'slash and burn' of the grass and reed jungles.[5] The great island called Majuli chiefly representing this *chapari* belt was created by the change of the course of the Brahmaputra towards the middle of the 18th century only.[6] Being intersected at several places by small streams, and surrounded by a number of smaller islands, this great island was, in fact, a cluster of islands—all of which "possess a stratum of rich soil above a deeper layer of sand and often clay.[7] However, the frequent changes of the river courses and occurrence of heavy floods made the riverine belt unsuitable for permanent cultivation. As a result, except for *pamuas* or temporary settlers[8] there were no permanent inhabitants in those areas including the smaller islands. The case of the great island Majuli, was however, different. Even before its creation in form of an island it was a strip of land on the south of the Luit or Brahmaputra, and the Ahom kings in the middle of the 17th century established big royal *Satras* there.[9] Still, a considerable part, possibly about one-forth of the area, remained undiscovered till at least the beginning of the present century.[10] The *Alamgirnama* mentions that the island now known as Majuli, lay to the south of the Brahmaputra and was in an excellent state of tillage. But the cultivated tract within it was bounded by thick forest which contained herds of wild animals including elephants.[11] In the entire *chapari* area were produced mustard seeds, *ahu* and *bao* rices, various pulses, pumpkins, gourds etc. by the temporary settlers from nearby villages.[12] It was because of this temporary nature that as late as the 19th century, for about two hundred miles below Sadiya, there was no town worth the name in the river bank areas.[13]

Beyond this *chapari* and riverine belt, lay the actual and major area of permanent settlements extending on both banks of the Brahmaputra to a certain distance till it reached the hill slopes. This belt was a broad one where the annual flood of the Brahmaputra and its tributaries abundantly deposited alluvium and made it suitable for permanent cultivation.[14] The average rainfall of about 75-150 inches per annum made the lands soft and clayey so that transplanted rice cultivation could be easily done here. According to the *Alamgirnama* "the country of Uttarcul which is at the northern side of the Brahmaputra is in the highest state of cultivation . . . (it) even surpasses Dacshincul in population and tillage".[15] The *Fathiya-i-Ibriya* also records that the north bank was more populous and abundantly cultivated.[16] But these reference to the north bank actually meant the northern region in Goalpara and Kamrupa and the western part of Darrang only through which the invading Mughal army in 1662 advanced. From Kaliabar onwards, most part of the south bank, as reported by them, was also under cultivation. It was, therefore, in this *rupit* belt that the plough using people had, in all probability, their earliest permanent settlements.[17] It was in this plain tracts that the earliest state of Pragjyotisha-Kamarupa, and in later times, smaller states of the Chutiyas, the Kacharis, the Koches, and the Ahoms with the productive utilisation of the soil, grew. During the early part of the Ahom occupation of the valley almost major part of the region except some selected areas, was under shifting cultivation pursued by the tribes mostly of Bodo Kachari origin.[18] After the establishment of the Ahom rule, the region was gradually brought under permanent rice cultivation. Early records show that the Ahom rulers in order to extend cultivation built large number of embankments to prevent frequent flood raids from the Brahmaputra and its innumerable tributaries,[19] and engaged people in gradually clearing jungle lands. As a result, in the middle of the 17th century, Shihabuddin Talish found that from Kaliabar in Nagaon to Gargaon, the Ahom capital, the entire region was full of gardens, orchards and extensive rice fields.[20]

Beyond this *rupit* belt and touching the foot hills regions was the submontane belt inhabited by tribes like the Bodo Kacharis, the Meches, the Garos and the Hajangs. Most of these tribes were under some chiefs with autonomous power under the Ahoms. This region being subject to heavy rainfall with an average of 80-120 inches per annum, was full of jungles and weeds. The early British records show that here too 'slash-and-burn' method of shifting cultivation mainly prevailed.[21] Beyond this region were living the hill tribes like Akas, Daflas, Miris, Nagas, Mikirs etc. The Ahom rulers had no direct control over their territories, but these tribes accepted very often Ahom suzerainty. Some of these tribes produced limited amount of rice in the hill slopes through terrace and shifting method. They used hoe and stick in stead of plough—a system which was till recent times prevalent among them.[22] As a result, economy of the plains producing large quantity of rice was inseparably linked up with that of the hills. For it was here that crops like cotton, ginger and other forest products including animal skin and fur were procured. Such hill products were exchanged for the products of the plains especially rice, salt and fish.[23] In our period of study, as discussed in a subsequent chapter of this work, both the plains and the hills regions were found to be interdependent in their economic life.

Thus the kingdom of Assam in its full grown stage under the Ahoms stretching over the entire plains of the Brahmaputra valley from Sadiya in the east to the eastern border of Goalpara in the west was under cultivation. That Assam was full of forests and jungles in the late 18th and the early 19th century, as recorded in some of the contemporary local and British records,[24] was due to some circumstances. Since the late 18th century to the close of the Ahom rule in the first quarter of the 19th century Assam passed through devastating civil wars and foreign aggressions so that almost half of the total population were killed and lands remained fallow allowing jungles to crop up.[25] It may be noted that jungles here grew in abundance within a short space of time due to its soil fertility and high rate of rainfall. It can therefore be very reasonably concluded that in spite of the best efforts of the Ahom rulers, agricultural extension was sometimes handicapped in the valley by natural as well as human factors. But the Ahom rulers fully utilised the existing manpower and every possible advanced technology to overcome them and thus brought considerable amount of land under cultivation even at the initial stage of their rule. It is to be mentioned that there was no limit to the process of bringing new lands under cultivation. The rulers always encouraged the subject population to go on encroaching the jungles for clearing new lands and adding to the existing ones. Such new lands were either made revenue-free or were given to the peasants for a lumpsum of its produce as revenue. This policy encouraged the peasants to open new lands time and again. The process being repeated vast areas of lands became level plain and fit for settled cultivation so that it was then registered as revenue paying land and distributed among *paik* peasants. Lands gifted to Brahmanas, *Pirs* or any religious institutions included, besides cultivable lands, a part of waste or jungle lands which were gradually being opened for cultivation. These were indeed the policies of the Ahom rulers to increase the amount of cultivable lands at the cost of the neighbouring forest and wastes. Thus it appears that in spite of all obstacles, substantial amount of cultivable lands were opened in Assam under the Ahoms. But unlike in Mughal India,[26] it was due to the lack of specific and sufficient contemporary records that in Assam it can not be said with certainty as to the exact amount of lands brought under cultivation under the Ahoms. Butler estimated in 1853 that only about 4.6 per cent of the total land of Assam was then under cultivation; he, however, admits that some centuries ago it was richly cultivated by an industrious and energetic people.[27] This is evidenced by the fact that travelling through the south bank of the valley above Guwahati, Shihabuddin Talish found that almost the entire region from Kaliabar to Gargaon and from the Brahmaputra to the southern hills was green with cultivation of paddy and fruits, flowers and orchards.

> "From Kaliabar to Garhgaon houses and orchards full of fruit trees stretch in an unbroken line, and on both sides of the road, shady

bamboo groves raise their heads to the sky. Many varieties of sweet scented wild and garden flowers bloom here, and from the rear of the bamboo groves up to the foot of the hills there are cultivated fields and gardens. From Lakhugarh to Garhgaon, also, there are roads, houses and farms in the same style; and a lofty and wide embanked road has been constructed up to Garhgaon for traffic.

In this country they make the surface of fields and gardens so level that the eye cannot find the least elevation in it to the extreme horizon".[28]

According to the *Alamgirnama*, the regions around Gargaon and above the Dhansiri, was marked with population and tillage and it presented on every side charming prospects of ploughed fields, harvests, gardens and groves.[29] The north bank below Guwahati according to this chronicle was more cultivated and largely populated.[30] Local chronicles bear ample testimony to the production of various crops in Assam in our period. It is found that beginning with the time of Sukapha, the Ahom rulers and their officers opened big paddy fields, i.e., farms known as *khats*[31] by employing thousands of *paiks*, slaves and servants. This was why king Rudra Singha (1696-1714) who collected a big force of about four lakh persons while planning to invade Bengal, could solely depend for its supply of food on domestic products. It is recorded that he kept aside Rs. 2,00,000 only to be expended for their salt, oil and other immediate necessaries.[32] It is to be noted that out of this force of four lakhs, three and half lakhs were native Assamese soldiers or *paiks* who might have cultivated at least seven lakh *puras* of lands at the rate of two *puras* allotted to each of them to be used in transplanted rice cultivation alone. Besides, they also grew other crops—both edible and cash, in additional lands acquired by clearing jungles. To this may be added the numerous *khats*, and *pams* (lands for temporary cultivation in riverine tracts) and assigned lands which were almost entirely put under cultivation. It is presumably because of such evidences that Hamilton observes that about ¾th of the total land area in upper Assam were under cultivation prior to the Moamariya rebellion.[33]

There is, however, no contemporary local record to estimate the exact volume of production of this period. Lack of large scale inter-state trade in paddy leads to further problem in its estimation. Still, if a plough of land (i.e., about 4-5 acres) could produce about 30 *vis* or 79 mounds of paddy, or 6 *vis* or 16 mounds of mustard,[34] the total out-put of these items can be presumed; and a family, as Talish reports, could collect at least a year's need in every agricultural season.[35]

Method of Cultivation

Agricultural Appliances

In north-eastern region, and as such in Assam, methods of cultivation mainly depended on ethnic heritage and geographical situation of the lands. Population in the north-east in medieval times primarily consisted of the tribes who were either in the process of Hinduisation or almost outside its pale. As a matter of fact, agriculture system as a whole here was largely marked by tribal ways of *farming*.[36] A small group of people in the valley in earlier times, and the Bhuyans and the Ahoms in subsequent periods only had a developed agricultural tradition and were associated with high yielding transplanted varieties of rice cultivation. The period therefore witnessed both tribal and non-tribal methods of cultivation, one age-old and less productive carried on mostly by the tribes and some of the commonality, and the other, comparatively modern and high yielding, carried on by the advanced sections, notably the Ahoms.

Sali rice which was associated mostly with the plains' people, particularly the Ahoms,[37] was a transplanted wet rice. Besides, it being a long maturing type, its cultivation also involved a lengthy process. Only such fields which could retain water or could be artificially irrigated from adjacent streams were reclaimed for its cultivation. A description of wet rice cultivation method as existed in this region till the 19th century is found in R. Brown's *Statistical Account of Manipur*, a method which is still followed in Assam. Brown says:

> "In June the rains having set in, the field is brought by successive ploughings and harrowings into a state of liquid mud, and in this the 'pung hul' (seed) is cast. The seed for the 'pung hul' is first quickened by being moistured with water and kept in a covered busket until it shoots. As this seed floats on the surface of the mud, it has to be carefully watched untill it takes root, and three or four leaves spring up, in order to protect it from wild duck and other birds. After this comes . . . transplanting. The seeds for the plants, which are destined to be transplanted are usually sown very close, in plots carefully prepared for the purpose. When the transplanting season arrives, the plants are pulled in handsful out of the ground; the roots are by washing divested of all the earth attached to them, and having been taken to the field, they are one by one separately inserted in the mud . . . they soon spring up and afford an excellent crop".[38]

Ahu is sometimes transplanted when it is known as *kharma*, but its production does not equal to that of the *sali*.[39] Some of the tribes of the submontane tracts and hills, e.g., the Bodo Kacharis, the Khasis, and the Jayantiyas had a slight different type of wet rice culture being done on the slopes and the plains of the hill areas. Here transplantation did not take place; seeds were sown thickly to provide ample space for the growth of the seedlings. The method which involved clearing of lands, levelling it, creating irrigation cannals to supply water from the hill streams, sometimes from long distances, weeding etc.[40] is labour consuming and similar to that of the sali cultivation. But this was a short maturing variety and known *asra aha*.[41]

As constant retainment of water in required amount in the field is a precondition for *sali* crop, levelling of the fields and making criss-cross ridges to divide the entire field into some small rectangular plots are very intensively done to serve this purpose. Shihabuddin Talish noticed that the Assamese make the surface of the fields and gardens so level that the eye cannot find the least elevation in it to the extreme horizon.[42] The aphorisms prevalent in popular circles in Assam and Bengal including present Bangladesh even till date and associated with the well known personality named Dak, make mention of making frequent ridges in the field, which will protect the paddy from getting dried in the month of *Ahin* and *Kati* (i.e., October and November) like the king who protects his queen from all dangers.[43] The aphorism even talk about the technique of different varieties of paddy cultivation including that of the *sali*.[44]

Almost all the tribes, Hinduised or otherwise, were adapted to shifting rice agriculture which included *ahu* and *bao*—the later being a long maturing variety. By this method with comparatively less utilisation of labour and use of very simple implements, rice and other crops were produced. A vivid description of this technique is found in John Butler's *A Sketch of Assam*:

> The natives set fire in the jungle to clear the land of cultivation, and to open the throughfares between the different villages and the aweful roar and rapidity with which the flame spread cannot be conceived. A space of many miles of grass jungles twenty feet high is cleared in a few hours. . . . The jungle is burnt down and for three successive years two crops are annually realised from it. In February, mustard seed is gathered in . . . and in June the spring rice sown broad cast is reaped. After the land has been thus improverished, it is allowed to remain fallow for three years and fresh jungle land is

prepared in the same primitive way and with most simple implements of husbandry.[45]

The description makes it clear that shifting methods of cultivation involves 'slash-and-burn' and 'sown-broad-cast'.

Ahu is a short maturing variety and by the beginning of summer, the crop is reaped. After that the land lies fallow for almost the next six to seven months. During this period, mustard and other *rabi* crops are grown there and by this method of double cropping the same plot of land can be utilised for different production.[46] Allen *et al* report that there is practically no rotation of crops apart from the system under which summer rice is followed by pulse or mustard, while pulse is usually sown on the rice seedling bed as it is thought to benefit the soil.[47] Records show that *ahu* seeds were sometimes mixed with those of *bao* while sowing broad-cast,[48] so that the *ahu* part would be reaped in summer and the *bao* in winter from the same plot of land. But this process of mixed cultivation was very limited.

Crops grown by shifting methods include among others mustard and pulses.[49] There are ample references to the method of slash and burn and sown broad cast in contemporary Assamese literature. For example, in the aphorism of Dak one finds that mustard seed should be very thickly sown and those of the pulses thinly in order to obtain the best results.[50]

The process of slash and burn makes the soil soft and involves simple implements. When the seeds are sown-broad-cast in the land thus opened, either plough or hoe is used. But the jhum crops carried on entirely by the tribes involve either "a short *dao* (long knife) fixed in a long handle, with which jungel is cleared, and a small hoe",[51] or a stick for making of some irregular holes in the soil wherein only seeds are planted.[52] While this technique of cultivation had been universal amongst all the tribes of the north-eastern region, the only difference was in the periods of allowing lands to remain fallow between production of two crops. For example, while the Garos left a piece of land uncultivated for seven years,[53] the Kacharis and the Karbis did it for three years,[54] and the Khasis and the Jayantiyas for two years.[55] In the riverine tract of the Brahmaputra valley lands were kept fallow even for eight to ten years[56] and in Arunachal Pradesh it varied from 3 to 10 years.[57] The period of fallowing lands depended upon certain factors, e.g., availability and fertility of lands, and density of population.

System of Irrigation

The system of irrigation essential for wet rice culture, was in some form prevalent since early times among some peoples of Assam which is evidenced by some contemporary foreign accounts. For example, Yuwan Chwang, the Chinese pilgrim, who visited the kingdom of Kamarupa in the middle of the 7th century noted in his account that water led from the rivers or from banked-up lakes flowed round the towns.[58] Minhas-ud-din Siraj refers to opening up of water dykes at the time of need.[59] Ralph Fitch, the English traveller, who visited the Koch kingdom (then lower Assam) during the second half of the 16th century, observed that the people used to innundate the lands by damming the streams above knee deep.[60] The *Katha Guru Charit* records how the Bhuyans of Assam built embankments on the rivers Tembuwani and Magurijan for protection of crops and houses.[61] It may be presumed that this land-lord class of medieval Assam a major section of whom were originally from outside the state, and who possessed advanced cultural traits, knew this method of agriculture well ahead of many of the indigenous tribes. Among the local people the Kacharis alone who had formed their state, knew how to irrigate lands.[62] It is found that they used to make dams across the hill streams and then led the stored up water to the fields through a net-work of dug up channels.[63] The Khasis and the Jayantiyas also adopted such system of water supply to the fields.[64] But it was only during the Ahom rule in Assam for long six hundred years (1228-1826) that irrigation and flood control work were undertaken as a state policy so that the flat level lands opened up for rice cultivation could contain only that amount of water which was actually needed for good cultivation. They gave much stress on this aspect of cultivation and constructed numerous spurs and embankments to prevent the flood of almost all the tributaries of the Brahmaputra in upper Assam.[65] For example, king Suhungmung became known as *Dihingia Raja* for his building an embankment on the Dihing river,[66] the largest of the southern tributaries of the Brahmaputra. One of his successors, king Pratap Singha tried to protect the lands by building large embankments, one of which was known as *Meragarh*[67], i.e., one which encompassed the entire land and protected it from any havoc of flood. This work of building spurs and embankments at strategic parts to control flood continued till the end of the Ahom rule. For example, it was during the reign of Kamaleswar Singha (1795-1811) that the river Bhogdai was excavated to supply water to the Jorhat town.[68] These embankments serving as preventive or protection measures to flood were also used as roads. Many of such embankments or *gar* are still in extent and are in use.

Method of Fertilising

Fertiliser in the modern sense of the term was only a recent development in Assam. Gunabhiram Barua in the last part of the 19th century remarks that the people of Assam did neither use fertiliser nor artificially irrigate their fields for the improvement of the fertility of the soil.[69] Chemical fertiliser was hardly in use in Assam even as late as the 20th century.[70] Natural vegetation and soil fertility made the use of manure in Assam a matter of secondary importance. Shahibuddin Talish noted in 1662 that here (i.e., in Assam) the soil is fertile, and whatever is planted grows well.[71] As pointed out by Robinson—the sense of temporary ownership of lands did not encourage the people to use manure and to improve its qualities.[72] Indeed, cultivation solely depended on what amount of fertility the soil naturally possessed. That would not however make much problem in the plains areas where soil was rich alluvial, but in the slopes of the hills and in the non-transplanted areas, the absence of manure affected produce. This is why large scale shifting technique was undertaken by both tribes and non-tribes and lands were kept fallow for years together to retain fertility.[73] The *Katha Guru Charit* and early British records mention how weeds had been burnt to clear the land by depositing ashes for *ahu* rice and mustard cultivation.[74] Both the hills and the plains tribes were adapted largely to this process. The Khasis and the Jayantiyas burnt the weeds and jungles to make the land fertile with their ashes.[75] It has already been mentioned that this method of manuring land in the tribal areas was noticed by John Butler. The most common class of manure was cow-dung, which was generally used to manure the vegetable gardens, plantain and sugarcane fields and in preparing the ground for growing seedlings for transplanted rice.[76] John M'Cosh has noticed that the Assamese prepared a well manured garden land to sow seeds to obtain healthy seedlings.[77] Butler, however, noticed that in Assam, excepting the fields close to the villages, the best lands were never manured.[78] The aphorism of Dak includes cow-dung, ashes, garbage from kitchen etc. as being used to fertile garden lands.[79]

Implements of Cultivation

Almost the entire population of the north-eastern region had only the simplest implements of cultivation. The inhabitants of the lower

Brahmaputra valley which came into contact with early Aryan migrants to this region, had possibly knowledge of the use of plough drawn by a pair of bullocks much earlier to those of the upper valley. This is indicated by the use of the term *hala*[80] to measure lands, i.e., lands which could be cultivated by one plough and a pair of bullocks. For example, Sankaradeva (1449-1569) the great Vaishnava reformer of Assam, who himself was a Bhuyan, inherited a landed estate along with thirty pairs of bullocks and twenty cows.[81] There was not much change in the set of agricultural implements when the British occupied this land in the early part of the 19th century. It is recorded in the *Gazetteers* that the agricultural implements included a wooden plough (*nangal*) with an iron tipped share (*phal*), wooden rakes (*jabaka*) and mallets (*dalimari*), a rough bamboo harrow (*mai*), yoke (*juwali*), sickles (*kachi*), bill hooks (*da*), knives and different types of bamboo baskets like *duli, mer, kharahi, pachi*, etc. to keep the products.[82] As we have noted earlier most of the tribes of the surrounding regions were throughout the period associated with hoe and stick. It is only recently that plough has entered to some of the tribal societies—notably among those of the plains. But the Ahoms had from the very beginning large scale use of plough, pulled both by bullocks and buffaloes. It is to be noted here that the Ahoms inherited the use of domestic buffalo in ploughing from their earlier abode in south-east Asia. It was, as pointed out by Gordon Luce, their use of buffalo that led to the growth of an enriched and settled civilization.[83]

Literary records of medieval times supported by early British records show that the extent of cultivation of the transplanted and high yielding rice in Assam was larger than that of *ahu* and *bao*, the figure being 70 per cent, 22 per cent and 8 per cent respectively.[84] Walter Hamilton also states that about ¾th of the produce of the whole crops formed transplanted rice.[85] Again, production of *sali* was much widespread in the upper Brahmaputra valley and that of *ahu* and *bao* or shifting rice, produced by slash and burn and 'sowing broad-cast' method had been extensive in the lower part of the valley.[86] Guha writes: "As one moves from the districts of Sibsagar, the cradle and core of the Ahom dominion in the eastern extremity of the valley towards the west or towards the north one finds that the importance of *sali* in the total rice crop goes on decreasing . . . the percentage of the total rice lands under *sali* in Sibsagar and Lakhimpur districts are the highest being 92 per cent and 85 per cent respectively".[87] The *Gazetteer* for the Brahmaputra valley also records that by the close of the last century there was not much *bao* and *ahu* in Sibsagar, Darrang and Lakhimpur.[88]

The account of the Mughal chronicler Shihabuddin Talish reveals that in upper Brahmaputra valley wet rice cultivation attracted the attention even of the foreigners. The absence of some significant extension of wet rice agriculture in lower Assam region can be inferred also from the account of Ralph Fitch who visited the Koch kingdom in 1585. He specially mentions a number of products like silk, musk, cloth, cotton, bamboo and cane grown in the area,[89] but makes no reference to rice cultivation. This implies that if there was any large scale cultivation of rice specially transplanted or wet, that was largely in the upper Assam region, the heart-land of the Ahom state only. The *Katha Guru Charit* which contains stores of information about the socio-economic life of the people of the lower Brahmaputra valley, frequently speaks of *ahu* rather than *sali* variety of rice.[90]

From the contemporary hills and plains relations it appears that although rice was cultivated in some parts of the hills also, it was done by the most primitive method and the production was very insufficient to meet the hill peoples' demand. This is why hunting, fishing and collection of wild roots and fruits were the major processes of maintaining their living in the hills in medieval times.

The apparatus and process of agriculture throughout the region, both hills and plains, remain till date almost the same except that chemical fertilizers and power tillers have come to be used in a very limited way in some rice fields. Shifting, jhuming, transplanting, ploughing, hoeing and sticking—all are in the same style or have only slight modifications. All types of *sali, ahu* and *bao* pursued with construction of dykes and embankments and manuring with burning weeds and mixing ashes with soil are still done depicting both the tribal and non-tribal heritage of the culture of the region.

Agricultural Crops: Volume of Production

Rice was the staple crop in Assam all through the ages as it is at present. Assamese literature make mention of varieties of rice grown in Assam. For example, the *Yogini Tantra,* an early 16th century Sanskrit work of Assam, speaks of as many as twenty varieties of rice, such as *Raja, Soma, Briha, Sighna, Banga, Raktasali, Keteki, Kalabinka, Narayana, Madhava, Pradipa, Ashoka*, and the like.[91] It also speaks of the spring time rice called *brihi* which evidently meant *ahu.*[92] The *Katha Guru Charit* also contains numerous references to rice cultivation and mentions its varieties.[93] Early epigraphic records mention that land measurement was done in ancient Assam in terms of the amount of rice (*dhanya*) produced in a given plot of land.[94] The *Fathiya-i-Ibriya*[95] and the *Alamgirnama*[96] both speak of extensive rice cultivation in Assam. Generally as mentioned elsewhere, rice was and is still of three main varieties—*sali, ahu* and *bao*. *Ahu* is usually sown broad-cast in the spring time and grown and harvested in summer. *Bao* which is also sown broad-cast, is a stemmed rice of long maturing variety. It is sown in low lying lands, usually in March-April and harvested in winter. Another kind of spring rice called *Baro dhan*[97] is sown (rather transplanted) in October/November and harvested in February/March. Of the *sali* there were two main varieties, of fine or *lahi* and coarse or *sali* or *bar*. Shihabuddin Talish states that there were no long and thin variety of rice in Assam[98] as produced in Mughal India such as *Sukhdas, Deozira Jinjin* etc.[99] which according to Abul Fazal were used in the royal kitchen of Akbar.[100] In spite of his statement, contemporary local literary records refer to various kinds of *bar* or big and *lahi* or thin and long varieties of rice as grown in Assam. Of those mentioned in the *Yogini Tantra*, most were of thinner variety. That various kinds of this variety of rice were grown in Asam in the period under discussion, can also be known from the *Statistical Account of Assam* by Hunter who has mentioned as many as eighty seven varieties of this rice.[101] Of these the most peculiar three varieties were *Chakua, Bara* and *Jaha*. Husked rice from *Chakua* and *Bara* were so soft that people in Assam used to take them unboiled just soaking them in water which was then called *komal chaul* or *bokachaul*.[102] It should be mentioned that in our period, the *paik* militia while in the battle field, mostly depended on *komal* or *boka-chaul* with ripe banana.[103]

The production of rice, particularly *sali* was voluminous, so that it was not only sufficient to meet the consumers' demands but had also sufficient surplus to meet any war time needs. Shihabuddin Talish noticed that during the year 1662 at Gargaon alone, the king possessed one hundred and seventy granaries, each containing one to ten thousand *mons* of paddy.[104] It should be mentioned here that the people of Assam who were solely dependent on paddy for their everyday needs treated rice as *lakhimi* or goddess of wealth. It was therefore not their custom to buy and sell it in the market.[105] The Mughal chronicler was astonished to mark this habit of the people here. The absence of rice in the list of export agricultural products to Bengal even as late as the 19th century[106] could be explained from this point of view.

Among other agricultural products the most important was oil seeds like mustard, which was grown in large quantities almost all over the state. This is evidenced by the early literary works and contemporary British records.[107] People used its oil for almost all domestic purposes—chiefly for lighting lamps in houses and temples rather than as an edible item. It, therefore, formed an important item of export to Bengal. Francis Hamilton estimated that about the beginning of the 19th century about 1500 *mons* of mustard seeds were exported to that state.[108] Sesame (*til*), another kind of oil seed, was also produced but in small quantity.[109] Among other cash crops, cotton was possibly the next of the most important. In the entire hill slope areas in both the north and the south banks of the Brahmaputra, and in the districts of lower Assam, cotton was produced.[110] The *Darrang Raj Vamsavali* mentions productions of cotton in the hills of Bhutan border.[111] Cotton formed a major item of export to Bengal. Francis Hamilton records that during the last days of Ahom rule about 7,000 *mons* of cotton were exported to Bengal alone bringing Rs. 35,000 to the royal treasury.[112] Besides, cotton had wide domestic use specially for making cloths. There was certain centres like Tantikuchi and Sualkuchi in lower Assam, where cotton cloths were largely produced. The Misings or the Miri tribe produced a special kind of cotton blanket known as *Mirijim*. In Assam a kind of paper also called *tulapat* was produced from cotton.[113] Weaving of cotton cloth in Assam was so widely prevalent that almost every family had a loom and Assamese people attained much perfection in the art of making cotton cloths. Along with cotton, other fiberous agricultural products like hemp called *san, rhea* (urtica nivea) and jute were cultivated which were mostly used for making fishing nets, ropes etc.[114]

Chronicles of Ahom period and other contemporary literature refer to cultivation of numerous pulses throughout Assam. These included *matimah* (phaseolus max), *mugmah* (Phaseclus minimus of Rumph), *kalamah* or *khechari* (Lathyrus sativus), *arahar* (eajanus caijan), *machur* (Lens esculenta), *butmah* (eicerarietmum), *lecheramah* (vigna sinensis), *urahi* (dolichos lab) and others.[115] Besides being an edible product *matimah* was also used as a substitute for plaster called *karal* in building palaces, temples, bridges, *maidams* or royal graveyards etc. The chronicle on engineering skill and technical sciences called *Changrung Phukanar Buranji* refers to large scale use of *matimah* for the purposes mentioned above.[116] *Mug, katari dabua* and other pulses were usually used as edible items.

According to Captain Welsh, there was cultivation of sugarcane almost in every part of Assam.[117] The *Fathiya-i-Ibriya*, mentions that there were three varieties of sugarcane in Assam—red, white and black.[118] Mohammed Qazim says that the sugarcane of Assam excelled in softness and sweetness.[119]

Vegetables and spices find mention in the contemporary literature as being cultivated in Assam from very early times. For example, the *Harsa Carita* of Banabhatta mentions that among the items presented to Harsavardhana by king Bhaskarvarmana of Kamarupa (594-650 A.D.) there were pumpkin gourds.[120] Early inscriptions of Assam also bear testimony to the cultivation of gourds. For example, the *Bargaon Grant* of king Ratnapala (1010-1040 A.D.) speaks of cultivable fields with clusters of gourds called *labukutikshetra.*[121] The *Yogini Tantra* mentions papaya and brinjal.[122] Some other vegetables are mentioned in the *Katha Guru Charit* and the *Dakar Bachan*. These include *porola sak*, *chuka sak* (Rumex vesiearious) *dhekia, babari, purai* (beescllaverbra) *lafa* (malva vesticiliata) etc.[123] Varieties of potatoes were produced both in the hills and the plains. A remote reference to the indigenous varieties of potatoes cultivated in the hill slopes of the region is found in the *Mahabharata* itself.[124] People of Assam till now use to cultivate a small variety of small potato called *aluguti* and it is very popular among them. They are also used to eating some wild varieties of tubers. Therefore, the potato introduced in the land by the British during the days of David Scott and first cultivated in the Khasi and Jayantiya hills, was of a different kind of larger size (*solanium tuberosum*) which is still differentiated in Assam as *Khasialu* from the indigenous variety. A very popular item of diet was *kharisha*, a kind of pickle, made from bamboo shoot.[125]

Among other crops grown in Assam, mention may be made of tobacco. The crop was introduced into India by the Portuguese during the reign of Akbar (1556-1605).[126] Very soon the Mughal aristocracy became addicted to the drug and cultivation of tobacco as a cash crop spread all over India.[127] In the mean time a Portuguese settlement was established at Rangamati in Goalpara in the early seventeenth century.[128] It is possible that after this, tobacco cultivation and its smoking became spread in Assam. Thus the production of the crop in Assam soon grew so that it could even be brought out for market. It is interesting to note that local sources point to the early existence of the crop in Assam. For example, some of the Assamese chronicles mention that a trader (*mudoi*) from Mughal Bengal bought tobacco in Assam secretly as early as the first quarter of the 17th century (about 1616 A.D.) and attempted to export it to Bengal.[129] In the middle of the 17th century Shihabuddin Talish found that tobacco was openly sold at Gargaon, the Ahom capital, at the rate of Rs. 3 per pipe.[130] It is to be noted that there was a place called 'Dhuwahat'[131] in the locality now covered by the island of Majuli within the present Jorhat district which indicates the existence of a market (*hat*) for *dhuwa, dhuwapat* or *dhapat*, i.e., tobacco. Be that as it may, tobacco production in Assam gradually grew. In the last part of the 18th century, Captain Welsh noticed that tobacco in Assam formed an important item of export. Like Mughal India, it also formed an item of aristocratic luxury in Assam. Assamese ballads associated with Maniram Dewan, the *Barbhandar Barua* of the last Ahom king Purandar Singha and a martyr of the 1857 revolt in Assam, speak of his *dhowakhowa*, i.e., hubble-bubble made of gold and silver.[132]

Poppy was also cultivated in Assam in some limited extent since long time past. According to S.K. Bhuyan the cultivation of poppy and the habit of its consumption in Assam might have been imported by the bordering tribes from China, where the production of the drug was prevalent since very early times.[133] During the period under discussion, opium was consumed in Assam mostly by the higher echelons of the society and it was also available in the market.[134] Captain Welsh found that king Gaurinath Singha was always in deep addiction with the drug.[135] A chronicle records that sometimes criminals were compelled to swallow opium tablet as a sort of punishment during the rule of the Ahoms.[136] It therefore appears that opium was locally produced in Assam during medieval period; but both its production and use—were limited. It was only after Assam's contact with Bengal towards the close of the 18th century that its consumption became wide-spread in the land,[137] so that when Captain Welsh was brooding over the prospects of British trade with Assam, he suggested to the Company Government to undertake the sale of opium in the land.[138]

Captain Welsh had also noticed that dying materials like indigo was cultivated in various parts of the country but in small quantity.[139] Other British records also make similar reference about the production of indigo.[140] Epigraphic records speak of a kind of land known as *nilmati*, a term which was presumably derived from the cultivation of indigo called in Assamese *nil*. Paintings in large number of manuscripts of medieval Assam point to its local production in the period under discussion. The *Katha Guru Charit* also mentions the use of indigo (*nil*) in dying clothes and yarn.[141]

Assam was, as it is now, famous for production of varieties of sweet and citrous fruits. Among the most popular and common fruit was plantain. It was not only offered as an item of offering to deities (prasad) and formed an indispensable item in ritual functions, but was also largely consumed by the public. The aphorisms of Dak describe in detail the methods of its plantation, selection of seedlings and use of fertilizer etc.[142] In Assam, besides the fruits, the plaintain tree itself was used for various social purposes. Both Dak and Tavernier speak of different uses of plantain trees.[143] In the account of Talish the production of *khar* or *kharani*, a kind of alkali from the plantain trees, is described, which was used as substitute for salt, the latter being very rare in medieval Assam.[144] Epigraphic sources speak of other fruits like jack-fruit, mango, *jamu, sriphala dimaru, bagari, amlakhi*, walnut etc.[145] The Chinese traveller Yuan Chwang while visiting this land found that jack fruits and coconuts were produced here in large number and were very popular.[146] The *Yogini Tantra* while detailing the list of fruits to be offered to the deity, makes mention of grape, plum, guava and pomegranate.[147] That large number of fruits were produced in Assam during our period can be affirmed by what was recorded by the visiting Mughal official chronicler Shihabuddin Talish, who found that mangoes in Assam were plentiful, sweet and free from fibers, but full of worms and of scanty juice. Here pineapples were very large, delicious and juicy.[148] Mohammed Qazim also noted the production of mangoes, plantains, jack fruit, pineapple, *panial* and *amlakhi* in Assam.[149] The *Alamgirnama* speaks very high about *panial*, which has much excellence of flavour so that it could be preferred even to plum.[150] During the middle of the 17th century, it was found that from Kaliabar to Gargaon houses and orchards were full of fruit trees stretching in unbroken line.[151] Stephen Cacella, the Portuguese traveller, who visited Koch Behar and Hajo in the first half of the 17th century, mentions that Koch Behar and lower Brahmaputra valley were famous for their fruits, especially mangoes, which were of better quality than those found in other parts of India.[152] Even Shihabuddin Talish remarks that oranges here are so plentiful that the people at Gargaon used to sell them at the rate of ten organes for one *paisa*.[153]

Among the citrous fruits cultivated in Assam, were lemon, *amara* (spondias Amara), tamarind, olive, shaddok, *amlakhi* (embica officinalia) *kardai* (averrhoa caranbola) *thekera* (*thaikol*) etc. The *Dakar Bachan* suggest various uses of the citrous fruits in the Assamese dietary systems.[154] Assam's *thaikol* fruits were in demand even outside Assam. Francis Hamilton while enlisting items of Assam's export to Bengal included *thekera* or *thaikol* amounting to 50 *mons* for Rs. 150 per annum.[155]

Many varieties of spices were cultivated and used in Assam in the period under discussion. Among these were the bay-leaf, black pepper, onion, ginger, garlic, coriandam, cardamom and many others. Eventhough chilly is said to have been brought to India only in the mid 18th century from South America,[156] early Assamese works mentions *jal* and *marich* in reference to preparation of delicious food.[157] It appears that the one which was brought from South America was no other than capsicum which was till then not grown in Mughal India,[158] hence was not found in Assam also. But here in Assam, *jalakia* both wild and cultivated, were probably in use since early times. Even now in the jungles of Assam some chillies are found and one of its varieties known as *bhot jalakia*, implying its association with the tribes on the Bhutan border, indicates its indigenous origin. There are other varieties called *khud jalakia* and *dhan jalakia* which may also be of indigenous origin. Among other spices, Mohammed Qazim mentions that Assam abundantly produced bay-leaf and pepper[159] which formed important articles of trade with Bengal. In Hamilton's list both black pepper and long pepper are found, about 100 *mons* of which were exported to Bengal annually for Rs. 800.[160]

Thus many agricultural products grew in Assam which were not usually found elsewhere in India. Shihabuddin Talish makes an interesting statement in this regard:

> Many kinds of odorous fruits and herbs of Bengal and Hindustan grow in Assam. We saw here certain varieties of flowers and fruits, both wild and cultivated, which are not to be found elsewhere in the whole of India. The coconut and *nim* trees are rare, but pepper, spikenard and many species of lemon are abundant.[161]

Development of gardening and cultivation of vegetables were systematically undertaken by the Ahom rulers. The noble class used to maintain well-knit gardens of both flower and vegetables during our period. The Mughal chronicler refers to the extremely elegant and fresh garden of the Barphukan at Guwahati 'round a very pure and sweet tank within the grounds of his mansion'.[162] Momai Tamuli who was the first incumbent to the office of the *Barbaruah* was said to have been an expert garden-keeper in his early life.[163] There was an officer called *Barichowa Barua* to look after the royal gardens.[164]

Gardening of different kinds of flowers, both wild and domestic, excluding possibly roses[165] were done by the people in Assam. Shihabuddin Talish recorded this in his account. The leaders of the Neo-Vaishnavite movement taught the people the art of keeping beautiful flower gardens where flowers like *jai, juti, malati, champa* and others were grown.[166] People of Assam were accustomed to chewing areca-nut and betal vine. The Ahom rulers stressed on maintaining a systematically planned areca-nut and betel-vine gardens in front of every house. Sources of our period confirm production and consumption of areca-nut and betel-vine.[167] The *Jayantiya Buranji* and the *Kamrupar Buranji* mention both ripe and unripe nuts chewn with two different varieties of vines.[168] It is believed by some that the term 'Guwahati' has been derived from areca-nut and betel-vine gardens existing here.[169] The *Ain-i-Akbari* refers to *pan* which was largely grown in Mughal Bengal,[170] now well-known as *Banglapatti*. Assam's betel-vine better known as *pan* of course differs from *Banglapan* and was used is small transactions.

In Assam, lac was produced in large quantities. It was through an insect that lac was extracted from a special kind of tree.[171] Lac produced by way of cutting stick from the same tree known as stick-lac, was an important item of commerce. Hamilton records that in the beginning of the 19th century about 10,000 *mons* of stick-lac were exported to Bengal and thus an amount of Rs. 35,000 was annually earned.[172] Assam also indigenously produced rubber. Indian rubber as it is known, was produced from a tree—*Ficus Indica*.[173] But its quality is inferior to those produced in America.[174] Thus Assamese chronicles mention that during the Ahom period rubber gardens were opened in the state by planting rubber trees.[175] There is still a place called *Athabari* in present Sibsagar district signifying rubber gardening in medieval period. Varieties of bamboos were cultivated by the people for "it was the only material with which the houses of the people were constructed" besides most other appliances and furniture of every day use.[176]

Sericulture and Other Agro-Manufactures

Making of silk thread and weaving of its cloth in Assam date back to the remote past. Early Indian literature contain reference to silk work in the Brahmaputra valley.[177] Both Kautilya and Banabhatta spoke highly of Assam silk.[178] Foreign travellers had also noted the sericulture work and silk products of Assam and its excellence. Pliny in the first century A.D. speaks of silk production carried on in Assam.[179] Tavernier has noted that silk in Assam was produced on trees and its stuffs were very excellent.[180] Ralph Fitch while visiting western Assam in the Koch days noticed its silk industries.[181] The writer of the *Alamgirnama* writes that the silk of Assam was very excellent, resembling that of China.[182] Sericulture as a profession on the basis of caste grew in Assam in the medieval period. The name *Katani*, which now stands for a caste-name, denoted the people who used to spin *pat* (mulberry) silk thread or weave cloth from it.[183] That silk production in the state was taken as a planned cultural pursuit, can be known from the fact that silk worm rearers and silk weavers were regularly taxed[184] in our period of discussion, and were categorised into specific *khels* or groups for assignment to the nobility including the king's household and the *Satradhikars* or religious heads to produce for them their necessary silk cloths.

There were three varieties of silk produced in Assam—*eri, muga* and *pat.* The *Katanis* were associated with *pat* silk, but *muga* and *eri* (sans erandi) were produced by almost all other castes except the Brahmanas. The *muga* silk obviously derived it's name from amber colour and hence it was characteristically Assamese.[185] *Muga*, cocoons are reared on trees like *adakuri* (tetranthera quedrifolin), *sowalo* (Tmacrophylla), *com* (machilus ordoratissima) etc. *Eri* worms are fed on castor plants (receinus eps) and *pat* worms on mulberry plants. All these trees and plants are still grown in Assam and worms are fed on their leaves. Robinson remarked that the same man reared the worm, spin the thread, wove the cloth and dyed it. The whole manufacturing process was therefore going on without much capital.[186]

The Ahom rulers imposed certain restrictions on the silk artisan class. Besides paying a higher rate of revenue in cash, they were also to give a part of their produce to the nobles and the kings. At the same time, the peasant *paiks* were also having certain limitations on their use.[187] Silk, particularly *muga,* however, formed an important item of export from Assam towards the later part of our period. This shows that the producer *paiks* had the duty of producing silk, but not the right of using it as they pleased. Francis Hamilton while giving the list of export items from Assam mentions that 85 *mons* of *muga* silk amounting to 10,350 rupees, and 75 *mons* of *muga* cloth for 17,500 rupees were exchanged for items from Bengal in the beginning of the 19th century.[188] Walter Hamilton in the last decade of the Ahom rule in Assam noticed that silk cloth formed a significant part of the clothings of the natives and a part of it was also exported.[189] That some of the Assam silk products entered the markets of Bengal and added to its export quantities and items could be known from Shihabuddin Talish also, who says that among other things, every year varieties of silk were exchanged for salt from Mughal India.[190] The fact that Qasimbazar in south-east Bengal alone supplied about 22,000 bales of silk for export which was almost equal to about 3.1 or 2.4 million pounds,[191] lends support to this view. Thus *muga* silk from Assam and Koch-Behar formed an important item of trade with Bengal, Bhutan, Tibet and in Coromandal and Malwar coasts.[192]

The most significant aspect of silk products in Assam was that it met to a great extent its peoples' demand for clothing. It was particularly from the *pat* and *muga* silk that the dresses of all upper classes were produced. Although in limited extent, the peasant class also used to manage their clothings from silk; these were usually of coarse variety, mainly *eri*.

Besides silk, cotton spinning and weaving also formed a very popular cottage industry in the period under review. Every woman of Assam from nobles to peasants excelled in this particular art. It is recorded that some of the Ahom queens tried to develop the art of weaving various cloths in medieval Assam. For example, queen Sarbeswari, wife of king Siva Singha, established in the capital at Rangpur a weaving school, where training in the art of weaving and designing cloths was given to the local girls.[193] She was also said to have collected styles and specimens of brockades and floral designs for borders of various clothes woven in the royal loom.[194] A professional caste known as *Tanti* (i.e., weaver) grew out of this profession. Traditions prevalent in Assam mention that it was Momai Tamuli Barbarua who first made it compulsory for every Assamese woman to spin a certain quantity of cotton thread every evening.[195] Ralph Fitch in the last quarter of the 16th century noticed its textile manufactures along with production of silk.[196] The noted Assamese Vaishnava reformer Sri Sankaradeva himself once acted as the headman of a professional weavers' guild at Tantikuchi in Kamrup.[197] He is also said to have produced a 60 × 30 yards large cotton cloth woven for the Koch king Naranarayan. This piece of cloth called *Brindavani Kapor* depicted scenes from the early life of Lord Krishna.[198] There are references to weaving beautiful cotton cloths in Narayanpur in upper Assam also.[199] Professional weavers were kept for making cloths for the royalty and the nobility, and some of them were also assigned to religious heads, priests or *Satradhikaras*.[200]

There are references to the fact that cotton cloths were beautifully embroidered with coloured threads or painted with various colours.[201] The arts of dying and painting developed in our period. Dying materials were prepared from some agro-products, such as, indigo, lac, *majathi* (Indian madder), and some natural herbs—such as *mishimi tita*, a kind of herb named after the hills tribe the Mishimis, leaves of *leteku tree, asokath*—a wild plant, turmeric, *thekera*—a wild fruit and others.[202] Besides bamboo musks, almost every manuscript was painted and illustrated. The most notable among them was the *Hastividyarnava*[203] by Sukumar Barkaith and the *Chitra Bhagavat* by Sankaradeva.[204] Cotton cloths were usually produced for domestic use; and there is no reference indicating its export to outside the state. But there were small village markets where cotton cloths were sold and some of the poor families earned their living by weaving and selling cotton cloths. *The Guru Charit* of Daityari Thakur refers to a weaver couple who earned about ten *pons* of (10 × 80) cowries (the price of which was equal to the ten *anas*) by this profession every day.[205] The Ahom kings had a large number of looms within the capital campus for production and supply of various kinds of cloths to the members of the royal household. Contemporary records show that a number of expert women weavers were engaged by the king to supervise the work.[206]

Two types of looms—one big, fixed on the ground with bamboo or wooden posts, and the other waist loom which was tied to the waist on one side, and to a post on the other. Waist looms which are still in use among some of the north-eastern tribes, are simple enough to handle as it takes very small space. Fixed looms called *tatsal* are popular among the plains people, particularly the professional weavers class called Tanti—the appliances of which are almost entirely made of bamboo. The Vaishnava biographical works and the Assamese chronides make mention of such *tat-sals* or domestic looms,[207] as used by the Assamese commoners in medieval times.

Agriculture based manufactures included wood crafting and bamboo works. The plough called *nangal* and its other implements—almost all were made from wood and bamboo. Boats cut out some times from a single piece of wood and sometimes by joining wood-planks served the multiple purposes of trade,[208] naval warfare and conveyance. The Ahom kings organised certain *khels* of expert artisans called *Naosaliya* for making boats.[209] Means of communication between distant lands almost entirely depended on boats. The houses of all the people from the king to the peasants were made of wood, bamboo and reeds.[210] Other implements used for agricultural purposes included *japi* (a kind of sun shade also used in rains), baskets, winnowing fans, mats, a kind of pole called *nal* used for measuring lands; *don* an appliance for measuring paddy and many bamboo implements for catching fishes. A class of expert craftsman called *Barhoi* were employed by the Ahom rulers to make wood and bamboo appliances. Many poor peasants exchanged such domestic appliances for rice etc. and earned their living.[211]

Different kinds of pottery and iron implements were used in agricultural works. Iron tipped share of plough, knife, a kind of bill-hook called *da*, hoe and axes were indispensable for clearing jungle lands. Hoes were also exported.[212] Besides, there were the sickle to cut paddy in times of harvest. All these were produced by a class of workers called *Kamar*, i.e., blacksmith.[213] The blacksmiths could earn their living by selling their products in village markets and some of them were attached to the royal houses, and the houses of the nobles and the *Satradhikaras*. Thus the kings and his nobles had their own blacksmithy *khels*, called as *Kamarsalia*. Pottery articles were largely used in agrarian life. *Paiks* engaged in this craft were called *Kumar*. Most of the pots manufactured by them were of a certain standard size so that the bigger one called *kalah* were used to measure some other agricultural items including jaggery, pulses, oil etc.[214]

Oil crushing and *gur* making are two important agriculture-based ancillary manufactures developed in Assam in the period of our discussion. Mustard oil was extracted almost by every family by an wooden appliance.[215] There was also the professional oil crushing caste called *Teli* who used a kind of cattle powered mill to extract oil. Assamese chronicles and biographical works of the Vaishnava saints refer to this caste. It is found that towards the later part of the medieval period, oil was also sold in the market, particularly in lower Assam.[216]

Gur or jaggery was largely used by the people throughout the state; but sugar manufacture is not recorded in any of the contemporary Assamese chronicles. The *Katha Guru Charit*, mentions that on his return from the first pilgrimage about the year 1493 Sankaradeva brought with him certain amount of sugar from other parts of India and gave it to his community and family members for taste.[217] The work contains certain other references to the use and manufacture of sugar in the Kamrup region in the subsequent period.[218] It, therefore, appears that while in strict exclusion from having contact with other parts of India, the peasants of upper Assam had no idea of the art of manufacturing sugar in our period, a section of the peasants in lower Assam due to their acquaintance with the systems that developed in Mughal India, had cultivated that art in Assam towards the late medieval period.[219]

From above, it appears that development of agriculture was considered as the most important aspect of the growth of the state in medieval Assam. Among agricultural crops, the most acclaimed and pursued one was transplanted *sali* paddy although side by side shifting and sowing broad-cast crops of *ahu* and *bao* were also cultivated. But *ahu* and *bao* crops were mainly confined to the tribes, and to the people cultivating in the low-lying and *chapari* lands, mostly in Kamrup. Rice was produced in surplus for the royal store but was not sold and exported;

it was chiefly expanded to meet the war time necessities. Of other crops almost all of them were produced for domestic consumption. Selected crops were cultivated in selected geographical zones and steps were taken to develop and protect the crops. Except a very few crops like mustard, cotton and silk, no stress was laid on production of cash crops. But then some of the produce of the soil having surplus over consumption were exchanged, chiefly for salt from Bengal. Agriculture, and manufactures based on it, provided almost everything of the daily needs of the peasantry except perhaps salt. This picture was however, a slight different in Kamrup where some peasants earned their living by selling their products including bamboo appliances, vegetables and fruits, or mortaged themselves to the rich peasants.[220]

Weights and Measures

Production and exchange of articles and collection of revenue in terms of produce necessitated certain units of weights and measurements in ancient and medieval Assam. Establishment of trade centres in the border areas, growth of small market places inside the state, and in later times, development of imports and exports specially with Bengal through the Kandahar choky led to the use of both bigger and smaller units to weigh and measure goods. While recording the export import items to and from Bengal the early British writers mention a unit called *mon* (*man/mound*) equal to forty *sers*.[221] The term was, however, in use in other parts of India including Bengal during the period under review.[222] It is, therefore, natural that the term *mon* might have been used in weight and measurement in the lower Brahmaputra valley, i.e., in Kamrup when it was under the Mughals. But the use of the term is found to be markedly absent in any of the contemporary records of the period in upper Assam. The lack of the use of the term *mon* in upper Assam denoting a unit of weight of solid articles was due largely to the lack of established trade and commerce in the region. Even the *Changrung Phukanar Buranji* which records in detail the amount of expenditure of each item while building the *maidams* and other buildings of our period, does not refer to this unit of measurement.[223]

The lowest unit to measure rice and such other commodities was *katha* which was about one half of a *ser*.[224] The term *chatak* was a more smaller unit used to measure smaller items like black pepper, spices etc. which was equal to 1/16 of a *ser* or ¼ of a *poa*.[225] The term *poa* was more popular than *katha* and was used even to count the number of *paiks* in medieval Assam, which usually meant ¼ of a *got* (a unit). In other occasions four *poas* formed a *ser*—a term which was largely used to weigh pulses, limes, powdered brick while using them for building purposes.[226] Examples of uses of different units of measurement for different articles in the middle of the 18th century may be had from some epigraphic records also. For example, the list of an estimated expenditure, to be incurred in connection with the annual worship of the deity in the Siddheswar temple in Kamrup is given in detail in the concerned land grant as follows.[227]

(a) For daily offering (*Naivedya*)

Name of Items	*Amount*	*Prices*
Pulses	2 *poas*	1 *ganda* 4 *karas*
Salt	1 *tola*	1 *ganda*
Ginger	1 *tola*	4 *karas*
Areca-nut	4 *ta* (i.e., pieces)	1 *ganda* 4 *karas*
Betel leaves	1 *guchi* (20 pieces)	1 *ganda* 4 *karas*
Oil (for lighting lamp)	2 *poas*	15 *gandas*
Ghee	1 *tola*	1 *ganda* 2 *karas*

(b) Expenditure for the *Durga Puja*

Name of Articles	*Amount*	*Prices*
Husked rice	40 *sers*	8 *anas*
Pulses (*mati*)	3 *sers*	15 *ganda*
Mug	2 *sers*	12 *anas*
Jaggery	2 *sers*	2 *anas*
Ginger	1 *poa* 1 *chatak*	2 *gandas* 2 *karas*
Areca-nut	5 *ta*	5 *karas*
Black pepper	4 *tolas*	8 *gandas*
Betel leaf	2 *bihas*	2 *anas* 10 *guchis*
Salt	½ *ser*	2 *anas*

(c) Expenditure for *Doul Yatra*

Name of Articles	*Amount*	*Prices*
Husked rice	40 *sers*	3 *anas*
Pulses (*mati*)	3 *sers*	15 *gandas*
Mug	3 *sers*	1 *ana*
Mustard oil	1 *ser*	1 *ana* 10 *gandas*
Salt	½ *ser*	2 *anas*
Jaggery	2 *sers*	2 *anas*
Milk	2 *bhars*	5 *gandas*
Betel leaf	1 *biha*	1 *ana* 5 *gandas*

Solid articles of small size such as areca-nuts, fruits and others were numerically counted. Units of counting such articles including those like needle were *ta, pon, kari, buri, ara, biha*, (20 pieces), *mutha* (2 bundles of 20 leaves each) *khila* (one piece), *ganda* etc. Four *karas* or cowries formed a *ganda* and twenty *gandas* formed *pon*, i.e., a *pon* of any fruit containing eighty pieces; and 16 *pons* formed a *kahan* or *kaon*. The same method of calculation was applied to count money. Four cowries formed a *ganda* or an *ana*, and sixteen *anas* formed a *taka* or one rupee.[228]

Articles which were otherwise weighed in terms of *ser, poa, adher* (half *ser*) etc. as stated above, were sometimes weighed or counted by certain units of measurement of unknown amount or quantity. For example, pulses like *matimah*, mustard oil, lime, powdered brick were weighed by a *dhol*, i.e., a hollow wooden drum and *kalah* (pot). The chronicles mention such units of measures as follows while recording the expenditure incurred in connection with the building of the royal palace at Gargaon in the year *Saka* 1680 (1758 A.D.); Jaggery—5432 *kalah* (big earthen pot), *matimah*—770 *dhol, san* 770 *dhol* ordinary lime—2670 *dhol*, refined lime—9750 *kalah*, fish 468 *dang* (weight that could be borne by two men by hanging the bunch of fishes in the middle of a strong bamboo pole), mustard oil—148 *dhol/kalah*, *dhuna* (resin)—1118 *ser*.[229] Sometimes the term *bhar*[230] was used to denote the weight that a man could carry. Other popular measures of weights were *don* and *pura* used for the measure of both land areas and grains or pulses.[231]

It appears that to weigh or count bigger amounts, units like *dhol*,[232] *kalah, bhar, dang, pura, don*, were used. But in case of smaller amounts, units like *ser, adher, poa, chatak, tola*, and *rati* (1/96th of a *tola*) were applied. Valuable items like gold and silver were weighed in terms of *tola* (1/80th part of a *ser*), or 180 grains troy and *maha*, i.e., ¼th of *tola*).

Use of units for smaller denominations indicate smaller volume of internal trade in the state. It is to be noted that while the bigger units—*dhol, dang, kalah* etc. were found to be used in upper Assam, the smaller units like *ser, adher, poa, chatak, kara, ganda* were found to be much popular in Kamrup area. Besides, in Kamrup rice was also sold, to which practice the people of upper Assam were not accustomed. This shows that everything produced from the soil including rice were brought and sold at least in the village markets in lower Assam while the same did not happen in the upper region of the valley. It is interesting that while the prices of the items expended for the functions and maintenance of the temple mentioned above are recorded in the epigraphs, the chronicle recording the same items as expended in building the royal palace at Gargaon and other buildings also, did not record the same. This shows that the growth of money economy in the lower Brahmaputra valley as a result of its early contact with the Mughals, had an important bearing on the development of its market system. But in upper Assam although volumes of agricultural crops were produced, these were mainly consumed at home or exchanged for the hill products, and a few items only like mustard were sold to the traders from lower Assam who in the later part of our period used to collect them from the houses of the peasants.

REFERENCES

1. Hadira, opposite to Goalpara was the last outpost of Assam in the west after the final battle with the Mughals fought in Itakhuli in 1682. See *CHAB*, p. 276. Gait in the basis of a map prepared by Rennell mentions that during the hey day of Ahom power, Manah formed the western boundary of the Ahoms on the north of the Brahmaputra and on south of that river it was some twenty one miles farther east, *HAG*, p. 189.
2. J.P. Wade, *A Geographical Sketch of Assam, 1800*, attached to his *An Account of Assam*, North Lakhimpur, 1927 (Reprinted from *The Asiatic Annual Register, 1805*), p. 3.
3. *DAR*, pp. 2f; John M'Cosh, *Topography of Assam*, Reprint, Delhi, 1975, p. 1. Gunabhiram Barua (*ABGB*, p. 1) mentions that excluding the hill areas, the total plains land covered about 17044 sq. miles. Also see *Mills Report*, Appendix-A, p. 56; Bhuyan, *op. cit*, p. 1.
4. *Asiatic Researches*, II, p. 131. At the time when Qazim writes of the Majuli island, it did not form an island in the strictest sense of the term. In fact, it was then a strip of land lying between the two rivers flowing westward in parallel, the Brahmaputra in the north and the Dihing in the south which Qazim wrongly puts as Dhonce (i.e., Dhansri). He, however, clearly writes that this Majuli was on the south of the Brahmaputra, implying that the great river unlike at present, did not form at that time the southern boundary of that island.
5. Powell, *Land Systems. . . .*, p. 418; *MEAG*, p. 10.
6. Wade, *Geographical Sketch*, p. 32; *ibid*. D. Nath, 'Majuli Dvip: Iyar Samaj Paribartanar Dhara', *Dibrugarh University Post-Graduate Journal*, 1990, pp. 43ff.
7. Wade, *Geographical Sketch*, p. 32.
8. Powell, *Land Systems . . .*, p. 396; *MEAG*, p. 10.
9. King Jayadhvaj Singha established four big *Satras* or Vaishnava monasteries in Assam, three of which were in this strip of land. These were Auniati, Dakshinpat and Garamur Satra. See *HAG*, p. 140.
10. Wade, *Geographical Sketch*, p. 33.
11. *Asiatic Researches*, II, p. 131.
12. Powell, *Land Systems . . . .*, p. 418.
13. *MEAG*, p. 10.

14. Allen *et al*, *op. cit*, p. 62.
15. *Asiatic Researches, II*, p. 131.
16. *HAG*, p. 141; Sarkar, *loc. cit.*
17. *MEAG*, p. 10.
18. Hodgson, *Eassay the First on the Koch, Bodo and Dhimal Tribes*, Calcutta, 1847, pp. 47ff; Dalton, *op. cit*, p. 82.
19. See for a detailed record of the rivers of Assam Wade's *Geographical Sketch.*
20. *HAG*, p. 141; Sarkar, *loc. cit.*
21. *AAH*, p. 73; Hodgson, *op. cit*, pp. 47, 154ff.
22. Butler, *Travels . . .*, pp. 157f.
23. *Ibid.*, p. 22; *DAR*, p. 357.
24. *ABGB*, p. 119; M'Cosh, *op. cit*, pp. 13, 29; Butler, *Travels . . .*, Appendix, p. 268 (see Table).
25. M'Cosh, *op. cit*, p. 13; *Buranji Vivek Ratna, II* (Ms); *Mills Report*, p. 621.
26. See for details, *Habib*, *op. cit*, pp.1ff.
27. The percentage of land under cultivation (i.e., 4.6 per cent) is deduced from the statistical table. See *Travels . . .*, Appendix, p. 268. According to M'Cosh about the middle of the fourth decade of the 19th century about 6/8 to 7/8 part of the valley lands were covered by dense jungles. See, *op. cit*, p. 13.
28. *HAG*, p. 141; Sarkar, *loc. cit.*
29. *Asiatic Researches*, II, p. 132.
30. *Ibid.*, p. 131.
31. See *Supra*, pp. 29ff.
32. Wade, *op. cit*, p. 133.
33. *AAH*, p. 26. For the Moamariya revolt see M. Neog, *Socio-Political Events in Assam Leading to the Militancy of the Mayamariya Vaisnavas*, Calcutta, 1982; D. Bora, *Moamariya Gana Abhyuttham* (As.), Guwahati, 1983; A. Guha, 'Neo-Vashnavism to insurgency: Peasants Uprising and Crisis of Feudalism in the Late 18th Century Assam', in A. Mitra (ed.), *The Truth Unites: Essays in Tribute to Samar Sen*, Calcutta, 1985; S.L. Barua, 'A Study of the Nature of the Moamariya Revolt', *JHR*, Vol. II, Dibrugarh University, 1974.
34. *AAH*, p. 36; W.W. Hunter, *A Statistical Account of Assam*, Vol. I, Reprint, Delhi, 1990, p. 47.
35. *HAG*, p. 150; Sarkar, *loc. cit.*
36. For technological development in India see J.B. Bhattacharjee, 'Technology, Social Formation and Historical Inquiry', Presidential Address, History, Archaeology and Culture Session, XVII Indian Social Science Congress, Bangalore, 1993; I. Habib, 'Presidential Address', Medieval India Section, *Proceeding of the Indian History Congress*, Varanasi, 1969.
37. Amalendu Guha in his paper 'Ahom Migration and Its Impact on the Rice Economy of Assam' (*Arthavijnan*, Vol. 9, No. 2, 1967) has established that the introduction of the *Sali* or transplanted wet rice in Assam may be particularly associated with the Ahoms who were traditionally associated with this culture (pp. 255f). It was Gordon Luce who forms the opinion that it was indeed in south-east Asia that the people first crossed the Rubicon between the dry and the wet rice—the latter, according to him, being "one of the great economic discoveries in the history of humanity". See 'Rice and Religion: A Study of Old Mon-Khemer Evolution and Culture', *Journal of Siam Society*, Vol. III, 2, 1965, p. 141; also D.G.E. Hall, *A History of South East Asia*, 4th edn, London, 1987, p. 237.

38. Calcutta, 1873, p. 87; also see Allen *et al*, *op. cit*, p. 62; C.A. Soppit, *An Historical and Descriptive Account of the Kachari Tribes of the North-Cachar Hills with Specimens of Tales and Folklore*, Shillong, 1885, p. 18.
39. Allen, *et al*, *op. cit*, p. 62; *MEAG*, p. 71.
40. Hunter, *op. cit*, Vol. I, p. 223; Allen *et al*, *op. cit*, pp. 489f.
41. *MEAG*, p. 71.
42. *HAG*, p. 141; Sarkar, *loc. cit.*
43. *Dakar Bachan* (ed.), H. Barman, Guwahati, 1977, Chapter VII, v. 13.
44. *Ibid.*, Chapter VII, vv. 18-19.
45. pp. 21f.
46. *Ibid.*, p. 23.
47. *Op. cit.*, p. 66.
48. *Ibid.* In Majuli island area some peasants even now follow the system in order to obtain a part of the crop before the summer flood submerges the land.
49. Powell, *Land Systems . . .*, p. 418; *KGC*, pp. 46, 59, 74.
50. *Dakar Bachan*, Ch. VII, v. 21.
51. Allen, *et al*, *op. cit*, p. 507.
52. Major A. Playfair, *The Garos*, Reprint, Guwahati, 1975, p. 34.
53. *Ibid.*
54. Butler, *Travels . . .*, p. 139.
55. Hunter, *op. cit*, I, p. 223.
56. Allen, *et al*, *op. cit*, p. 62.
57. R.C. Dutta (ed.), *Resarun*, Vol. VIII, No. 2, 1982, p. 3.
58. Cf. P.C. Choudhury, *op. cit*, p. 334.
59. *Tabaqat-i-Nasiri*, I, p. 565; K.L. Barua, *op. cit*, p. 151.
60. J.H. Ryley (ed.), *Ralph Fitch: England's Pioneer to India, Burma etc.*, London, 1899, p. 112; William Foster, *Early Travels in India* (1533-1619), Oxford, 1921, p. 25.
61. *KGC*, pp. 11f, 38.
62. Butler, *op. cit*, p. 23.
63. *Ibid.*
64. P.R.T. Gurdon, *The Khasis*, Reprint, Delhi, 1975, p. 40; Hunter, *op. cit*, II, p. 223.
65. See *HAG*, p. 267.
66. *ABS*, p. 17; *ABHB*, p. 40; *HAG*, p. 86.
67. *Ibid.*, p. 27; *Ibid.*
68. *TB*, p. 150.
69. *ABGB*, p. 187.
70. Allen *et al*, pp. 65f.
71. *HAG*, p. 144; Sarkar, *loc. cit.*
72. *DAR*, p. 217.
73. Butler, *op. cit*, p. 23.
74. *KGC*, pp. 94f; Allen *et al*, *op. cit*, p. 66.
75. Gurdon, *op. cit*, p. 4; Hunter, *op. cit*, II, p. 223.
76. M'Cosh, *op. cit*, p. 29; Allen *et al*, *op. cit*, p. 65.
77. *Op. cit*, p. 29.
78. *Op. cit*, p. 22.
79. *Dakar Bachan*, Ch. VII, vv. 24-25.
80. See the *CP Inscription of the Kalika Temple at Jayantiyapur, Saka* 1710 (1788 A.D.), *PSN*, Ins. No. 89; also see Ins. Nos. 98, 99.
81. Sankaradeva, *Bhagavata* Canto VI, vv. 5534-35.

82. Allen, *et al*, *op. cit*, p. 63.
83. F. Thomas, *op. cit*, p. 137; Gordon Luce, *loc. cit*; Hall, *op. cit*, pp. 237ff; *The Cambridge Economic History*, I, p. 482.
84. Allen *et al*, *op. cit*, p. 65.
85. *A Geographical, Statistical and Historical Account of Hindostan and the Adjacent Countries*, Vol. 2, Reprint, Delhi, 1971, p. 742.
86. Allen *et al*, *op. cit*, p. 65.
87. *MEAG*, p. 71.
88. Allen *et al*, *op. cit*, p. 65.
89. Ryley, *op. cit*, p. 112; Foster, *op. cit*, p. 25.
90. pp. 26, 56, 59, 63, 75, 101, 311.
91. Chapter V, vv. 289-91.
92. *Ibid.*, v. 287.
93. These include *Bar, Lahi, Pharma, Phaphari, Bara, Jaha, Bao, Mathanga, Maigutiya*, etc., p. 518.
94. For a detail records for almost all such available grants mentioning and accounting amounts of rice produced as ways of land measurement see *JHR* iv, pp. 49ff.
95. *HAG*, p. 142; Sarkar, *loc. cit.*
96. *Asiatic Researches*, II, p. 131.
97. The term *Bodo dhan* might have been derived from the fact that this variety of rice is very popular among the Bodo tribe of Assam.
98. *HAG*, p. 142; Sarkar, *loc. cit.*
99. *Ain-i-Akbari* (tr.), H. Blochmann, Vol. I, 3rd edn, Calcutta, 1977, p. 60; P.N. Chopra, *Some Aspects of Society and Culture During Mughal Age*, 2nd edn, Agra, 1963, p. 35.
100. *Ain-i-Akbari*, I, p. 60.
101. Vol. I, pp. 369f.
102. S.K. Bhuyan, *Studies in the History of Assam*, Guwahati, 1985, p. 147.
103. *Ibid.*
104. *HAG*, p. 134; Sarkar, *loc. cit.* In Assam one *mon* or mound in the later part of the 19th century, was equal to 40 *Sers*. Se Hunter, *op. cit*, I, p. 195; also see *Infra*, p. 219.
105. *HAG*, p. 150; Sarkar, *loc. cit.*
106. *AAH*, p. 46.
107. *KGC*, p. 87, 94; *AAH*, p. 58; *Report of Captain Welsh* in Mackenzie, *op. cit*, p. 388; M'Cosh, *op. cit*, pp. 29f; Walter Hamilton, *op. cit*, p. 742. Allen *et al* noticed that about the close of the 19th century some more than a quarter and a million acres of land were under mustard cultivation. See, *op. cit*, p. 64.
108. *AAH*, p. 46.
109. *Ibid.*, p. 58.
110. Ralph Fitch speaks of Bengal cotton as the most famous in contemporary India. See Ryley, *op. cit*, pp. 25, 28; Foster, *op. cit*, pp. 112, 118; *ABGB*, p. 188.
111. vv. 53-54.
112. *AAH*, p. 46.
113. S.L. Barua, 'Agriculture in Assam in the Ahom Age', *JARS*, 1976, p. 59; *KGC*, p. 36.
114. *DAR*, p. 67.
115. *Jayantiya Buranji*, p. 102; *PSN*, Ins. No. 150; *DAR*, pp. 90f.
116. See *Changrung Phukanar Buranji* (ed.), B.K. Barua, Dibrugarh University, 1992, pp. 3ff.

117. See Mackenzie, *op. cit*, p. 387.
118. *HAG*, p. 142; Sarkar, *loc. cit.*
119. *Asiatic Researches*, II, p. 131.
120. *Harsacharita* (ed.), E.B. Cowell and P.W. Thomas, London, 1897, pp. 212f; P.C. Choudhury, *op. cit*, p. 335.
121. See lines 52-55; M.M. Sharma, *op. cit*, pp. 159, 165.
122. Section II, Chapter V, vv. 277-79, 316-19.
123. *KCG*, pp. 79f; *Dakar Bachan*, VIII, v. 10.
124. Cf. S.K. Chatterji, *Kirata-Jana-Kriti*, 2nd edn, Calcutta, 1974, p. 31.
125. *Jayantiya Buranji*, (ed.), S.K. Bhuyan, 2nd edn, Guwahati, 1964, pp. 98ff, 123.
126. I. Habib, 'Technology and Economy of Mughal India', *Indian Economic and Social History Review*, Vol. XVII (I), p. 4.
127. *Habib (B)*, p. 48.
128. Bhuyan, *op. cit*, p. 51.
129. *DAB*, p. 110; *PAB*, p. 63.
130. Cf. J.N. Sarkar, *The Life of Mir Jumla: The General of Auranzeb*, 2nd edn, New Delhi, 1979, p. 322; Sarkar, *op. cit.*
131. *KGC*, p. 65. It is recorded that Madhavadeva (1489-1598), the disciple of Sankaradeva, discarded smoking of tobacco and *bhang* or true-hemp as unethical to the tenants of neo-Vaishnavism, see *KGC*, p. 55. If this is believed then there is a possibility of the existence of a local variety of tobacco in Assam even before its introduction by the Portuguese.
132. *Maniram Dewanar Git* as prevalent among the folk-people of Assam.
133. S.K. Bhuyan, *Asam Buranjir Katha*, Guwahati, 1989, p. 92. Maniram Dewan in his *Buranji Vivek Ratna, II* (Ms) mentions that Langi Panisia, the first incumbent to the post of Barphukan, first imported opium to Assam from the west (*Bhati*).
134. Shihabuddin Talish records that opium was sold in Gargaon at the rate of Rs. 16 per *tola*. See *HAG*, p. 137; Sarkar, *loc. cit*; also Mackenzie, *op. cit*, p. 388.
135. *Letters of Captain Welsh to Cornowallis*, dated 4th Dec. 1792, and 1st July 1793.
136. *AB*, p. 271; Bhuyan, *Asam Buranjis Katha*, p. 93.
137. *Buranji Vivek Ratna*, II (Ms); Martin, *op. cit*, p. 671; *ABGB*, p. 168.
138. *Letter of Welsh to Cornowallis*, dated 4th Dec., 1792.
139. Mackenzie, *op. cit*, p. 388.
140. *DAR*, p. 234.
141. p. 208.
142. Chapter VII, vv. 22-26.
143. *Ibid.*, v. 23; V. Ball, *Travels in India*, Vol. II, 1925, p. 283.
144. *HAG*, p. 146; Sarkar, *loc. cit*, *Khar* or *Kharnai* i.e., alkali was also produced from husk of the *matimah* by burning them to ashes. In the *KGC* there is a reference to the production of *khar* by this process. See p. 300.
145. Cf. B.K. Barua, *op. cit*, p. 95.
146. S. Beal, (tr.), *Buddhist Records of the Western World, (Si Yu Ki)*, Vol. II, London, 1905, pp. 195f; P.C. Choudhury, *op. cit*, p. 184.
147. Section II, Chapter 5, vv. 277-79, 316-19.
148. *HAG*, p. 142; Sarkar, *loc. cit.*
149. *Asiatic Researches*, II, p. 131.
150. *Ibid.*
151. *HAG*, p. 141; Sarkar, *loc. cit.*
152. C. Wessels, S.J., *Early Jesuit Travellers in Central Asia (1603-1701)*, The Hague, 1924, p. 128.

153. Cf. *Habib (B)*, p. 101fn.
154. Chapter VIII, vv. 19, 22, 23.
155. *AAH*, p. 46.
156. J.N. Sarkar, *Mughal Economy: Organisation and Working*, Calcutta, 1987, p. 19.
157. *KGC*, p. 254; *Dakar Bachan*, VIII, vv. 8, 11.
158. *Habib (B)*, pp. 49f.
159. *Asiatic Researches*, II, p. 131; *AAH*, p. 46.
160. *AAH*, p. 46; given in Appendix-A.
161. *HAG*, p. 142; Sarkar, *loc. cit.*
162. *Ibid.*, p. 141; *Ibid.*
163. B. Sharma, *Durbin*, Guwahati, 1961, pp. 1ff; S.K. Bhuyan, *Lachit Barphukan* (Ass), 2nd print, Guwahati, 1980, p. 15.
164. *Durbin*, pp. 1ff; *KGC*, p. 188.
165. H. Dhekialphukan, *Asam Buranji*, Guwahati, 1829, p. 106.
166. Sankardeva's *Kirtan Ghosha*, Guwahati, 1983; *Haramohan* Section, vv. 24-26.
167. *Harsacarita*, pp. 212f; *Asiatic Researches*, II, p. 131; *HAG*, p. 147; M.M. Sarma, *op. cit*, pp. 183ff; Sarkar, *op. cit.*
168. *Jayantiya Buranji*, pp. 98f; *Kamrupar Buranji*, p. 86.
169. S.N. Sarma, *A Socio-Economic and Cultural History of Medieval Assam*, Guwahati, 1989, p. 75.
170. Vol. I, pp. 80f, 391, 416, 449.
171. The Garos produced lac on *arahar* plant and the Assamese on a kind of ficus, like *maj*. See Gohain, *op. cit*, p. 203.
172. *AAH*, p. 46.
173. M'Cosh, *op. cit*, p. 34.
174. *Ibid.*
175. *SAB*, p. 218.
176. *DAR*, pp. 49f; *The Comprehensive History of Assam*, Vol. III, p. 111.
177. See B.K. Barua, *op. cit*, pp. 103f.
178. *Arthasastra*, Book II, Chapter XI, vv. 80-81; Cowell, *op. cit*, pp. 211f; P.C. Choudhury, *op. cit*, pp. 241f; B.K. Barua, *op. cit*, p. 103.
179. W.H. Scoff, *The Periplus of the Erythrean Sea*, London, 1912, p. 267; P.C. Choudhury, *op. cit*, p. 340.
180. V. Ball, *op. cit*, p. 281; B.K. Barua, *op. cit*, pp. 103f; P.C. Choudhury, *op. cit*, p. 343.
181. Ryley, *op. cit*, p. 112.
182. *Asiatic Researches*, II, pp. 132f.
183. *Buranji Vivek Ratna*, II (Ms); *ABGB*, p. 179; P.N. Gohain Barua, *Asamar Buranji*, 19th edn., Guwahati, 1937, p. 4.
184. See *Infra*, p. 170.
185. *DAR*, p. 227; B.K. Barua, *op. cit*, p. 104.
186. *DAR*, p. 227.
187. *Ibid.*, p. 233; *ADB*, pp. 419ff; Dhekialphukan, *op. cit*, p. 107.
188. *AAH*, p. 45.
189. *Op. cit*, p. 75.
190. *HAG*, p. 144; Sarkar, *loc. cit.*
191. Cf. *Habib (B)*, p. 55.
192. Bhuyan, *op. cit*, pp. 52f.
193. Cf. Bhuyan, *Studies in the History of Assam*, p. 70.

194. *Ibid.*
195. *CHAB*, p. 438; Gohain, *op. cit*, p. 154.
196. Ryley, *op. cit*, p. 112.
197. *KGC*, pp. 213ff.
198. *Ibid.*, pp. 208ff.
199. *Mahapurusha Sri Sankaradev Aru Sri Madhavdeva Charitra* by Daityari Thakur, (ed.), H.N. Dutta Barua, Nalbari, (1978), v. 271; *KGC*, p. 68.
200. *DRV*, vv. 550-51; *KGC*, p. 251; *PSN*, Ins. Nos. 31, 82.
201. See *KGC*, pp. 208ff.
202. W.A.M. Duncan, *Dye and Dying in Assam*, London, 1895, pp. 22f.
203. It is a work on elephantry written by Sukumar Barkaith during king Siva Singha (1714-1744), now published from the Department of Historical and Antiquarian Studies, Guwahati.
204. It is the *Bhagavata X* by Sri Sankaradeva. For other aspects see M. Neog, *Sankaradeva and His Times*, Guwahati, 1965, pp. 307ff.
205. vv. 1250-1251. The *KGC* (p. 393) mentions that such a couple demanded one pon of cowri per cubit of cloth woven.
206. *KGC*, p. 251; also see *ABHB*, p. 128f.
207. *KGC*, p. 68, 208ff, 213ff, 251, 393; Daityari Thakwe's *Guru Charit*, vv. 1250-51; Bhuyan, *Studies in the History of Assam*, p. 70.
208. Commercial boats were of big size of about 60 ft. long and 6-7 ft. wide made out of wood planks which lasted for more than ten years. Gohain, *op. cit*, p. 202.
209. At the top of the Department of boat-building was the *Nao-saliya Phukan* who supervised the affair of boat-building and repairing in the royal manufactories (*naosals*) at ten places like Rangpur, Gargaon, Majuli, Dergaon and others. He had under him 6-7 thousand *paiks*. See Wade, *op. cit*, Introduction, p. XXXV; *ABGB*, p. 208.
210. *Asiatic Researches*, II, p. 134.
211. *KGC*, p. 331.
212. *AAH*, p. 46.
213. *Ibid.*, p. 63; Artisans of various classes were sometimes gifted to the religious heads or the *Satradhikars* for production and supply of articles for the uses of the Satra-inhabitants. See, *PSN*, Ins. No. 82. See Appendix-C.
214. See *Chang-rung Phukanar Buranji*, pp. 6ff.
215. Dhekial Phukan, *op. cit*, p. 100; *AAH*, p. 63.
216. For example, the *CP Land Grant Inscription of Siddheswari Temple, Saka*, 1665 (1743 A.D.) (*PSN*, Ins. No. 24) mentions the price of mustard oil as 1½ anna per *ser*.
217. *KGC*, p. 33.
218. *Ibid.*, pp. 133, 159, 225, 275, 360, 379 etc.
219. Such sugar producers were known as *Chenisaliya*. See Dhekial Phukan, *op. cit*, p. 104.
220. *KGC*, pp. 307, 331, 542, 545.
221. *AAH*, pp. 45f; Hunter, *op. cit*, I, p. 195.
222. See *Habib (B)*, pp. 391ff.
223. This chronicle gives the measure in terms of *dhol, kalah, gat, dang*, etc.
224. See Hunter, *op. cit*, I, p. 133.
225. *PSN*, Ins. No. 24; Hunter, *op. cit*, I, p. 195.
226. *Changrung Phukanar Buranji*, p. 2f.
227. *PSN*, Ins. No. 24; for the addend part of the inscription see p. 177. Also see Appendix-B.

228. Hunter, *op. cit*, I, pp. 133, 195.
229. *Changrung Phukanar Buranji*, pp. 6ff.
230. *PSN*, Ins. No. 137. For the word *bhar* see *Infra*, p. 225.
231. 1 *pura* = 3 or 4 *dons*, and 1 *don* = 5 *sers* of grain or 3 *sers* of paddy; also see Hunter, *op. cit*, I, p. 133.
232. The *dhol* is taken as equivalent to 3 *puras* or 9 or 12 *dons*; see, Hunter, *op. cit*, I, p. 133.

4

REVENUE ADMINISTRATION

Methods of Land Revenue Assessment: The Population Census

Land revenue formed the basis of the state's existence in the medieval period. In a society of self-sufficient economy, where money hardly played any serious role in state-maintenance, land, labour and the produce of the soil formed the major sources of meeting the state-expenditure. Rulers of medieval Assam, therefore, took every care to register quality-lands and the peasantry who worked on it.

The Ahom rulers had a peculiar system of collection of revenue. Here not only the soil, but also its people were considered as property of the state.[1] It was in terms of labour in stead of cash that a peasant had usually to pay his revenue. The rulers also used to pay the state officers their salary in form of land assignments and by way of allotting workers—both *paiks* or peasants and domestic servants.[2] Every able-bodied individual of the age group 16-50 years was considered as revenue paying unit so that larger the number of peasant-population, higher was the estimation of revenue amount. It was because of this fact that census of adult male population of the state was very strictly taken so that every working man would be registered for the state's service. It appears that at the initial stage of the Ahom rule, population census was not very systematically undertaken. But in course of time, the kings became very particular about it and tried to increase the number of population through various ways like subjugation of surrounding tribes, assimilation achieved through inter-marriage with them and by the process of Ahomisation.[3] By the 16th century the kingdom grew larger and it needed expansion of the state machinery also. It should be mentioned that for an alien power in a new situation, the existence and upbringing of the state lay in the organisation of its military strength. The recurring inroads of the neighbouring hill tribes and the attacks by the Turko-Mughal powers of Bengal from the west, necessitated a strong army and a substantial economy. This resulted in the development of a well-knit population census of the state and their compulsory engagement in the defence and production purposes.

From the *buranjis* it appears that population census was undertaken as an important state affair even since the time of the founder king Sukapha. The *Deodhai Asam Buranji*, for example, records that Sukapha was accompanied by nine thousand people.[4] In another occasion, the chronicle records that on his arrival at Khamjang on the Patkai range, Sukapha found that he lost a few of his fellow companions in course of his campaign led across the jungles of the Patkai hills, whereas some new ones from among the subdued had joined him. He therefore instructed the *pandits* to record the number of people meanwhile lost including those newly gained.[5] This shows that records of numerical strength of the population of the state was maintained since its inception. This became gradually a serious state affair along with the growth of the state's demand for higher amount of revenue and greater need for cultivation. From what is recorded in the chronicles it appears that king Subinpha (1281-1293 A.D.) was the first Ahom ruler to make a census of the people in order to allot them to work under the great *Gohains.*[6] It appears that the process was carried on regularly by the succeeding Ahom rulers. Chronicles have recorded that king Suhungmung *alias* Dihingia Raja undertook a serious census operation in the state in the beginning of the 16th century.[7] In this connection Gait writes that in 1510 "an enquiry was made into the number, condition, and distribution of the people and they were divided into clans",[8] and then one from every family was made to serve the state.[9] As recorded in the chronicles, king Pratap Singha brought further improvement to the system. It was he, who introduced a regular census by personally supervising the operations and arranged people into some specific work divisions called *khels.*[10] Pratap Singha then created the post of a chief executive officer known as *Barbarua* and engaged its first incumbent Momai Tamuli to arrange and organise the peasantry called *paiks* in certain order. This arrangement of *paiks* into an institutionalised form came to be known as the *Paik system.*[11] Prior to this, one man from every family served the state, although some families had 4-6 working men,[12] whereas possibly there were some which had only one. To do away with this disproportionate demand, Momai Tamuli resettled the old villages by breaking big families and established new ones,[13] and arranged the *paiks* into *khels* making it compulsory that one member of every *got* of four *paiks* was to render service to the state.[14] Thus the present district of Nagaon derives its name from villages newly established by Momai Tamuli Barbarua.[15] In such villages members of different communities, castes and professional groups, like Brahmana, Ganak, Kayastha, Kalita, Koch, Keot, Chutiya, Buruk, Baria, Ahom, Garia, Maria, Dom, Hari were made to live.[16] Historians, while studying the agrarian structure of Mughal India, point out that such a mixture of population provided basic background for a self-sufficient economy.[17] Subsequently, however, this pattern of village settlement might have proved ineffective in Assam due to difficulties in immediate mobilisation of *paiks* in times of need, and in context of growth of money-economy and market system as in other parts of India[18] towards the later part of our period so that it was changed and the *paik* peasants were settled according to their professions and castes to make their mobilisation for a particular task easy.

The results of the census were properly recorded in registers called *paikar piyalar kakat*.[19] A class of officer known as *Kakatis* were appointed for recording all matters and preserving the documents for the royal archives. So much care was taken to make the census list complete and accurate that not an able bodied-man was left unregistered nor classification of *paiks* remained incomplete, nor the name of any foreigner could get into the list. The *buranjis* mention that the system of physical verification, classification and settlement of *paiks* was carried on with such accuracy that the state faced practically no problem in estimating and collecting the amount of revenue due to it in the form of personal service.[20]

It would, however, not be feasible to ascertain the exact number of *paiks* during the entire period of Ahom rule in Assam due to paucity of sufficient records. Only this much can be said that during at least the first three hundred years of the Ahom rule in Assam till a major territorial expansion of the state by Suhungmung Dihingia Raja was achieved in the first half of the 16th century, the region was sparsely populated. As mentioned elsewhere, king Sukapha was accompanied by about nine thousand men to Assam. From the chronicles it can be gathered that the number of the Marans, the Barahis and the Kacharis in the region of the extreme south-east of the Brahmaputra valley, where Sukapha laid the foundation of his state, was about sixteen thousand souls.[21] The numerical strength considerably increased during the time of Suhungmung, who not only conquered and annexed the entire Chatiya kingdom but also conquered a part of the Kachari territory and added to his kingdom the population of Habung and the Bhuyan territories in the south and the north-west.[22] Gradually, because of a prosperous and stable economy, population of the state normally began increasing. By the time the kingdom attained its zenith in the last quarter of the 17th century, it rose to some lakhs. The official records of emperor Aurangzeb contains that Assam's infantry and cavalry forçes in the war of 1670-71 consisted of some one lakh souls.[23] The chronicle translated by Wade, records that out of a total of four lakhs of *paik* soldiers collected by king Rudra Singha to make a campaign against Bengal in 1714 A.D., *kanri paiks* or foot-soldiers collected from his own kingdom numbered 2.6 lakhs excluding about one lakh *apaikan chamuas*[24], i.e. those *paiks* who did not work as foot soldier. Supposing that every *paik* or *chamua* had a family of an average of six members, the total number would rise to more than twenty lakhs,[25] excluding entirely, of course, the nobles and the aristocrats whose number as computed by Guha, consisted of about one per cent of the total population of the state.[26] Gunabhiram Barua estimates the total number to be 24 lakhs during the time of Rajeswar Singha[27] out of which some nine per cent may be roughly taken as consisting of slaves, servants and other bondsmen, and excluding the one per cent top aristocracy, the rest may be taken as the total working and non-working *paiks* (*apaikan chamuas*) whose strength was something about sixty five per cent and twenty five per cent respectively.[28] On the basis of such a calculation, the division of population of Assam during the mid-17th century may be shown as follows:

Aristocracy 1%	*Slaves & Bondsmen* 9%	*Chamua* 25%	*Kanri Paik* 65%	*Total* 100%
24,000	2,16,000	6,00,000	15,60,000	24,00,000

Methods and Amounts of Collection

(i) *Revenue in Form of Manual Labour:* At the initial stage of the system, a *paik* was a manual worker. But later on, as a result of the expansion of the state machinery and growth of different professions, non-manual working groups also became necessary.[29] As a result, the working force of the population was categorised into certain groups, such as, revenue paying in the form of manual labour, i.e., the *kanri-paiks* and the *paikan-chamuas* or the artisan class, and those who did not have to render manual service to the state, i.e., *apaikan chamuas*. It was then that the Ahom rulers made provisions for best utilization of the state lands. While they gave to every *apaikan chamua* a minimum of seven *puras* of best quality *rupit* lands to maintain his status, they allotted to the *paiks* through their clans or *khels* (a commune of a group of working *paiks*) certain amount of such lands at the rate of 2 *puras* (about 2.66 acres) per individual *paik.*[30] Each individual *paik* against his share of the allotted lands was to serve the state for a period of three or four months in a year. If best quality *rupit* lands were not available for distribution, proportionate amount of inferior lands viz., twice, thrice or four times (*dviguni, triguni chouguni*) of the *rupit* lands were given.[31] This state service was requisitioned in the form of working in the fields of the king, the members of the aristocracy and all other nobles including the religious heads of the *Satras*, priests, *pandits* and junior state officers. The *paiks* were also engaged in building embankments, roads, clearing jungles, constructing houses and royal palaces, besides acting as soldiers in times of war.[32] Since there was heavy demand for working—man for all these purposes, there was none left recruited as *paik* and hence there was also none capable of working who was landless. This implied that every peasant had cultivable land for his maintenance and that every piece of suitable land was under cultivation. W.W. Hunter while preparing the *Statistical Accounts of Assam* during 1871-1874/5 found that it was not customary for the people of Assam to leave lands uncultivated.[33] Such a situation naturally leads to the conclusion that with the object of collecting more revenue in form of man-power, the rulers took all possible measures to increase the amount of high yielding cultivable *rupit* lands by cleaning jungles.

(ii) *Revenue in Kind and Cash:* There was the system of paying revenue in cash also. A *paik* peasant, on rare occasions could avail the option to pay revenue in cash at the rate of Rs. 2 or 3 per head instead of giving his personal service.[34] But monetisation of Assam's economy was coming very late. Money therefore became hardly collected at the hands of the peasants. It should be noted that till at least the second half of the 18th century, money circulation and its production in Assam was very limited.[35] One chronicle mentions that when the Koch king Parikshit Narayan was captured and brought to Delhi by the Mughal army in 1612 A.D., he made a request to his ally, the Ahom king Pratap Singha, to pay on his behalf an amount of three lakhs of rupees to the Mughal emperor Jahangir to get him released. But Pratap Singha having no substantial deposit of money in the state treasury, had to decline the request. As mentioned in the chronicle, it was only after this incident that king Pratap Singha began imposing certain taxes to be paid in cash.[36] As a matter of fact, it can be very reasonably presumed that peasants in Assam at least till the 17th century could hardly pay revenue in cash.

Paiks of good birth wealth and rank, who were classed as *chamua paiks* or simply *chamuas*, could claim exemption from personal state service by paying a commutation money[37] or by giving certain portion of their produce or by rendering non-manual service to the state.[38] In the chapter on land systems we have mentioned that every family, besides having the *rupit* lands, possessed on hereditary basis a homestead plot which provided for its garden for fruits and vegetables. Although such lands were given as revenue-free in earlier periods, the civil war in the form of the Moamariya rebellion and the need for payment of the standing army that was formed thereafter, needed more revenue, so that the rulers had no alternative but to impose a cash revenue on the homestead plots also "as an equivalent for the exemption of these lands from assessment".[39] This new tax was collected in cash in Kamrup region as house tax or *ghar-kar*,[40] in Darrang as hearth tax or *juhal-kar* or *charu-kar* and in eastern Assam as *ga-dhan*, i.e., poll tax at the rate of Re. 1 per head,[41] or a tax in kind equivalent to 5-8 annas per *bigha* of land.[42]

The Ahom administration allowed *paiks* to reclaim waste lands by clearing jungles. This resulted in the opening up of additional lands known as *obar* or *katanimati* (i.e., land opened by clearing jungles) over and above the regularly allotted two *puras* of *rupit* lands by those whose man-power permitted this. Lands of this sort which could be enjoyed by a *paik* earlier as revenue-free until they got registered in the next census,[43] also became liable to taxation. The later Ahom rulers imposed a tax of Re. 1 per *pura* (*Katanimatiadhan*) on those lands also.[44] In the *chapari* belt or the inundated regions, emigrating peasants reclaimed lands to grow temporary crops. For this, the peasant was to pay a plough tax usually in terms of his produce.[45] The tribes who used the hilly lands for cultivation of cotton and other jhum crops without using plough within the territorial limit of the kingdom of Assam were to pay a tax known as hoe tax.[46] From the nature of trade-exchanges carried on between the hills and the plains,[47] and the tributes paid by the hill tribes to their Assam over-lord, it becomes clear that this hoe tax was also taken in terms of produce. Mills has recorded, therefore, that the revenue of the state was realised in most cases in kind or produce and in terms of manual labour.[48] The chronicler of Aurangzeb, Shihabuddin Talish mentions that if the kingdom had been administered like the Mughal empire, i.e., if cash revenue would have been collected, about forty to fortyfive lakhs of rupees would have been collected annually from Assam.[49] This shows that revenue was till then usually collected in terms of labour and produce.

Revenue in terms of labour was not much difficult to collect. As we have mentioned earlier, a register recording the actual number of *paiks* was maintained by the *Kakatis*, and there were strict instructions to the officers from the *Bara* at the bottom to the *Phukans* or the *Baruas* at the top to mobilise the *paiks* under their control for state services. The system was so complete that no *paik* was left unregistered and every officer mobilised his *paiks* in a short notice.[50] The artisan class whose labour the state had directly requisioned were also organised into *khels* and each *kheldar* or the officer of the *khel* was made responsible for obtaining their service to the state for a specific period.[51] *Paiks* of such artisan classes were permanently attached to the state establishments and were made to serve for the whole or a part of the year, in their specialised professions or for the whole season in which a particular work became feasible. For example, while the iron-smelters (called *Lo-saliyas*) had to smelt iron during the entire winter season,[52] the boat-makers (*Nao-saliyas*) and the goldsmiths (*Sonaris*) attached to the royal establishments called *sal*,[53] were to engage themselves in their service for the whole year in rotation as per *paik*-system.[54] It was the duty of the *Majindar Barua* to supervise the entire matter of revenue collection, issue papers with royal seal when a land grant was made and keep record of revenue paying *paiks*[55] on behalf of the *Barbarua* in upper Assam, who was the chief executive of the state. It usually happened that the artisan classes, instead of paying a cash revenue at certain rates, as mentioned earlier, used to pay it in terms of their produce, which they personally deposited to the immediate officers who finally deposited it to the royal treasury. Whatever cash money was collected, it was done by the *Tekelas* or constable working under an officer called *Tekela Bara*. The *Katha Guru Charit* records how a *Bara* insisted upon collection of the royal share of revenue from the *paiks* under his control.[56] Thus at the lower level, the *Baras, Saikias* and the *Tekelas*, and in the upper level the *Kakatis* or the record keepers and the *Bharali Barua*, the *Majindar Barua* and finally the *Barbarua* were the officers associated with the process of revenue collection both in cash and kind and their receipts and payments.

(iii) *Revenue in Lower Assam:* Revenue administration in lower Assam, which was under the direct supervision of the *Barphukan*, was somewhat different from that of upper Assam. This was due to the fact that a part of lower Assam which previously formed the kingdom of Koch-Hajo as we have mentioned earlier, was for a considerable period under the Mughal jurisdiction. When the Ahoms annexed this part of the country, there was already an well established system of administration in Mughal pattern which was almost impossible to be replaced by a new one entirely. Under the Mughals, the entire kingdom of Koch-Hajo was divided into four provinces or *sarkars* known as Dhekeri, Bangalbhum, Kamrup and Dhakhinkul.[57] For convenience of collection of revenue each *sarkar* was divided into some *parganas*; a *pargana* into some *taluks*.[58] While the overall supervision of all the *sarkars* was performed by the *Faujdar*, stationed at Guwahati, a *pargana* was placed under the supervision of a *Choudhury* who in fact was a *zamindar*.[59] When the system of collecting revenue by the *Kroris* was abolished, it were the *Choudhuris* who collected revenue from the peasants and for this service, they received as commission a certain portion of the revenue collected, and also some revenue-free lands as assignments.[60] Gradually his office became hereditary and his descendants acted as *zamindars* in their acquired territory.[61] The *Choudhury* was assisted by a number of revenue officers like *Patwari, Talukdar* (in charge of the *Taluk*), *Thakuria, Patgiri* and *Mena*. These officers also were allowed to hold certain amount of revenue-free lands as remuneration for their duty as revenue collectors and a part of the peasants' labour or their produce.[62]

On the charge of revenue collection there was in every village, a small officer either a *Thakuria* or *Patgiri*, who was assisted by a secretary called *Gaon Kakati* or *Kholar Kakati*, whose duty it was to calculate the number of *paiks*, the amount of produce to be paid as revenue and the extent and nature of service to be offered to the state.[63] The revenue officers deposited the collection with the *Choudhury*. The account of this collection was maintained by one *Patwari*.[64] The entire affair of collection and deposition of the whole region was maintained by the *Barkayastha* and the *Bujarbarua*.[65] The land grant epigraphs of medieval period frequently mentions these officers as associated with land and revenue.[66] They were all responsible to the Barphukan, who was at the helm of the affairs. John M'Cosh estimates that the average revenue thus collected from a *taluk* was about Rs. 500 per annum.[67] But the rate of collection might have varied from *taluk* to *taluk*, or from *pargana* to *pargana*. A copper plate inscription recording the event of conferring the charge of the Choudhuryship of the Bausi pargana in the year *Saka* 1738 (1816 A.D.) during the reign of Chandrakanta Singha on a certain person records the fixation of its annual collection at Rs. 1,300 per annum only.[68]

The Ahom rulers while retaining the Mughal pattern of administration in Kamrup at the time of its occupation by them had retained the system of collecting cash revenue instead of personal service of the peasants like that under the *khel* system.[69] They made another alteration. This was that although like in upper Assam, land allotment per *paik* was retained at two *puras*,[70] his land revenue (*jaigiri-dhan*) was raised to Rs. 5 instead of Rs. 2 as in upper Assam.[71] Peasants cultivating in waste lands had to pay a lumpsum amount as revenue in terms of cowries.[72] Half of the entire amount was collected in cash and the other half in terms of produce.[73] Since the profit of the *Choudhuris* who were in charge of overall revenue administration of the *parganas*, consisted of the difference between the rent which they collected and the revenue that they paid to the state, they also used to receive presents and extra amounts from the peasants. As a matter of fact, the *Choudhuris* and their subordinates collected from the peasants as much revenue as they could; but they paid to the royal treasury, presumably on one pretext or other only 2/5th of what they actually collected, which according to Francis Hamilton was equal to Rs. 32,000 per annum in the beginning of the 19th century.[74] The result was that like Mughal India the intermediaries imposed heavy burden of exactions over the peasants so that the latter in order to get rid of the oppression left their hearth and home and entered either the company's territory or the upper Assam region for relief.[75] S.K. Bhuyan has pointed out that the extent of oppression of the *Choudhuris* depended largely on the personality of the Barphukan at Guwahati whose weakness could encourage a *Choudhury* to collect more amount than the actual due.[76] This was why the post of the Barphukan sometimes became so lucrative a job that person like Badan Chandra during the period of turmoil paid a bribe of Rs. 60,000 to the Barkandaz leader Hazari Sing to get this office for himself.[77]

The autonomous tribal territories numbering more than fifteen in all ruled by their own chiefs but under the protection of the Ahom rulers included those of Rani, Luki, Beltola, Panbari and others. They maintained their own revenue systems, most presumably of tribal character.[78] These chiefs except the Raja of Rani paid to the Ahom king tributes, mostly in terms of goods and produce, rather than in cash. From the chronicles it appears that the chiefs besides paying some tributes annually, also paid on certain occasions, as in the times of war or on their visit to the Ahom king, some additional amount in form of man-power and goods. The latter included ornaments of gold, silver and brass articles, cotton cloths of various sizes and other hill products, including elephants, horses etc.[79] The *Deodhai Asam Buranji* mentions that the Kachari king Tamradhvaj even offered an area of land on the mouth of the Yamuna river in Nagaon along with forty people to serve king Pratap Singha.[80] The chief of Rani according to Francis Hamilton, was to furnish 621 *paiks* for the king annually and present articles worth Rs. 5,000.[81]

One of the tributary chiefs or *Rajas*, the most well-known and respected, was the Koch king (*Raja*) of Darrang whose territory was administered in line with Kamrup. The *Raja* was to give as revenue 6000 *paiks* to the Barphukan at Guwahati per annum.[82] These *paiks* were to act as soldiers and labourers under the Barphukan on the king's behalf.[83] They were also to give a hoe and some rice each, amounting roughly to rupees three thousand per annum.[84] Besides, there was also the system of cash collection from them by an officer of the rank of a *Barua*, at the rate of Re. ¼ per *paik*. Duties were levied at the rate of Re. ¼ to ½ on every cow or bullock sold in Darrang.[85] Other *paiks* were to pay taxes in gold, cowtails and blankets of Bhutan,[86] besides the usual hearth tax of Re. 1 per family. Moreover the state was surveyed and resurveyed since the time of Rudra Singha to ascertain the number of *paiks* so that maximum amount of revenue could be realised.[87] All these resulted in the increase of tax burdens on the peasants of Darrang which resulted subsequently in the people's protest against the Ahom monarchy during the reign of king Lakshmi Singha.[88] For that reason towards the close of the 18th century king Gaurinath Singha was made to agree to receive from the Darrang Raja a cash amount of Rs. 50,000 per annum instead of demanding *paik* labour and other taxes.[89] From what has been mentioned above, it appears that the Darrang Raja himself had to collect more revenue from the people in order to pay to the Assam king and to meet his own expenditure. When such was the demand from the Raja himself, the intermediaries like the *Barua, Rajputra* or the prince, the *Saharia*[90] the *Hazarika, Saikia, Bara, Buruk*[91] and *Duaria* (customs officer) exacted from the people such an amount that it would meet all their expenditures besides adding to the profit of these collectors.

Land Survey and Measurement

The other method of revenue assessment was to survey and measure lands, whereby lands were classified into certain categories,[92] and their amount for distribution among the *paiks* was determined whereby the extent of work-force was ascertained. As elsewhere mentioned, the Ahom kings were from the very inception, concerned with land distribution along with men at least amongst the great nobles. As a matter of fact, king Pratap Singha who first systematised the population distribution and settlement of villages, introduced also a regular system of land survey and measurement.[93] The next such survey seems to have been undertaken during the reign of Chakradhvaj Singha (1663-1671),[94] who personally supervised the work of preparation for a war against the Mughals and tried his best to find out the actual number of working-men in the state[95] by ascertaining the amount of cultivable lands to be distributed among them. During the subsequent periods land survey and resurvey were regularly undertaken by almost every ruler of Assam with the twin objectives of utilisation of lands and development of agriculture. But the *modus operandi* of the survey and the units of land measurement cannot be known for certain due to lack of proper records. From what is known, it can be said that the system differed from what was in vogue in early Assam but had some similarity with that of the Mughals.

The process of systematic and regular land survey and measurement under the Ahoms started since the beginning of the 17th century. One chronicle mentions that king Pratap Singha introduced the system of measuring lands by using a bamboo pole which was 7½ cubits and 4 fingers long,[96] a system which was used in measurement of land in Mughal Bengal.[97] The process might have been confined to limited areas and was not used in the entire state. The work, however, remained suspended during the period of the Mughal wars, which covered the major part of the 17th century. After the end of these wars, it was resumed by Gadadhar Singha. It is said that during the period of his concealment, when he was moving in disguise also in Kamrup, he saw there the Mughal system of land survey, and being convinced of its utility, he introduced the same in Assam after his accession to the throne.[98] The process was started in Sibsagar during this king's reign and was completed for the whole kingdom during the rule of his successors by the middle of the 18th century.[99] Gait says that this method of survey was in vogue till the time of the British, when they first occupied the land, i.e., the area of each field was calculated by measuring the four sides with a *nal* or bamboo pole, 12 ft. long.[100] The Ahom kings therefore imported a number of expert surveyors from Bengal to carry on the work. Thus a thorough land survey as in Mughal India, was undertaken and was completed for the whole state during the next one century.

In medieval Assam different units of land-measurement denoting different measures of lands were in use. Epigraphs of the period make mention of *puti* or *putaka*[101] as a unit of land measurement. Alongside, the Assamese chronicles make mention of another term—*pura*.[102] The term *pura* as a unit of land measurement was also found in a number of the inscriptions[103] of the period under discussion. It should be mentioned that *paiks* were allotted lands always in terms of the measure—*pura*. While the area of lands denoted by the unit *putaka* is not much clear, modern scholars have identified it as equal or equivalent to a *pura*[104] of four standard *bighas* of lands of 14,400 sq.ft. each. Besides *pura*, lands were also measured in terms of *hal* or *halisa* and *locha* or *lecha*.[105] While *hal* or *halisa* stood for an area of land that could be cultivated by one plough at a time which was considered equal to one standard *bigha*,[106] a *locha* was a much smaller unit.[107] Three inscriptions of medieval period recording the gift of lands to the Kalika temple of Jayantiyapur, mention *hala* and *kedara*.[108] The term *kedara* can not be properly defined due to lack of auxiliary references. It is possible that these terms were used to denote particular measurements of land in Assam before the Mughal system was imitated. Subsequently, however, lands were measured as per certain units and methods prevalent in Mughal India,[109] and the term *putaka* or *pura* were treated as equal to four *halas* or four standard Bengal *bighas*.

After the innovation in the land survey, the length and breadth of a piece of land was measured by a bamboo pole varying from 7-8 cubits in length. A copper plate inscription recording the settlement of a dispute over the Bardowa Satra in Nagaon, dated *Saka* 1721 (1799 A.D.) mentions a pole of 7 cubits in length.[110] Another inscription of *Saka* 1725 (1803 A.D.) mentions a pole 8 cubits in length.[111] Yet another inscription found at Beltala in Guwahati, mentions the pole to have been in length 7 cubits, 1 span and 4 fingers.[112] A chronicle mentions that while surveying lands, king Pratap Singha used poles which were 7½ cubits and 4 fingers in length, bounded by silver rings in both ends.[113] Another chronicle speaks of a measuring pole called *tar*, 7 cubits long, one measure of which was called a *beon*.[114] The term *beon* was also found to be used in measuring building sites during the rule of the Ahoms.[115] Thus it appears that two different poles of varying length were used in measuring lands in our period. The land-records, however, duly recorded the length of the pole by which a piece of land was measured. In any case the average length of a *nal* or a pole may be taken as almost equal to 8 cubits or 12 ft. which again was equivalent to one *beon* recorded in inscription[116] and chronicles. While taking the statistics of Assam, Hunter found that before 1852 land in Assam was measured by a *tar* or measuring rod of 7 *hats* (cubits) 1 bist (span) and 4 *angulis* (finger breadths) in length equal to nearly 11½ English feet,[117] the two ends of which might have been tipped with silver rings so that the length of the registered rod could not be tempered by anybody. It is interesting to note that the Mughal surveyors also used a sort of pole, the two ends of which were permanently marked by iron rings.[118] That this method of land measurement was similar to that used in Mughal Bengal can be learnt from certain epigraphic and literary records. The *Kachari Buranji* mentions a plot of land measuring 24 *puras* and 10 *lochas* which was 114 *beon* in length and 85 *beon* in breadth.[119] The *Copper Plate Land Grant Inscription of the Brahmachari Satra*, dated *Saka* 1722 (1800 A.D.)[120] also mentions the measure of a plot of land which was calculated as per the method prevalent in Mughal India where the area was determined by multiplying the mean length by its mean breadth.[121]

The new measurement system helped in proportionate or equal distribution of land among the *paiks* as well as among the nobles and officers depending upon their respective ranks. This indirectly necessitated the accurate determination of the number of *paiks* in the state and formed the basis on which land revenue was ascertained.

Lands were not only surveyed and measured, but their records were also maintained and officers were appointed to man the department. Persons expertised in the field of land survey and measurement were imported and engaged to undertake the work and to train the local officers. It is found that officers of the rank of a *Barua*, known as *Darabdhara Barua* superintended the work of survey and measurement.[122] Thus the entire lands, occupied and cultivable, excluding homestead and jungles,[123] were surveyed and their areas were determined and recorded. The papers recording these were known as *perakagaz* or *perakakat*[124] (literally means paper preserved carefully inside wooden boxes called *pera*). Even Sankaradeva, the great Vaishnava reformer of medieval period of Assam, mentions *perakakat*.[125] The *perakakats* were not very durable. Therefore, important records, particularly of the revenue-free lands gifted to the temples, *Satras*, mosques or to the Brahmanas and the *pandits* were preserved in a permanent form by engraving them on copper plates called *tamrapatra* or *phali*.[126] While the original copper plate was kept by the donee, two paper copies (*kakat*) of the grant were preserved in the royal archives.[127] In case no copper plate could be given to the donee, and he got the entire set of *perakakat* lost or brittled, fresh duplicate copies of the *perakakats* were issued to him.[128] For our period of study we have a number of *tamrapatras* which are still preserved.[129]

While surveying lands almost all the important revenue officers had to be present. For instance, land grant inscriptions show that when a piece of land in a particular area was gifted to a Brahmana or a temple, officers like *Barua, Barkayastha, Choudhury, Patwari, Talukdar* and *Thakuria* in Kamrup; and the *Rajputra, Saharia, Hazarika, Saikia* and *Bara* in Darrang were duly informed by the donor[130] who were required to be present on the occasion.

Revenue from Artisan Classes

Artisan *paiks* were not to serve the state with manual labour like the *kanri paiks*. The state therefore imposed a higher rate of taxes on them. Thus the goldsmiths (*Sonari*) and the bell-metal workers (*Kahar*) had to pay Rs. 5 each per annum, the oilmen (*Teli*) and the fishermen (*Kaivarta*) Rs. 3 and the silk worm rearers (*Katanis*) Rs. 2 each[131] besides giving a portion of their produce as presents to their officers.[132] In most of the cases, however the term of money payment was realised in kind so that the artisans paid their entire dues in terms of their produce.[133] Some of the artisans of particular professions like goldsmith were engaged to work for the king for the whole year.[134] All the artisans like the *kanri paiks* received from the king revenue-free lands at the rate of 2 *puras* each.[135] The *Sonowals* or the gold-washers who extracted gold from the sands of the rivers numbered some 10,000—12,000 in the later part of the 17th century, and each of them had to deposit one *tola* of gold per year to the royal treasury,[136] besides an extra charge of 3 *ratis* for first class, 4 *ratis* for medium class and 6 *ratis* for ordinary grade of gold extracted.[137] The officers like the *Barua, Bhandar Kakati, Phukan Tekela* and *Bara*, associated with the entire process of collection each received certain amount of allowances: one *tola* in every 20 *tolas* of gold collected by the *Phukans* and the *Baruas*, half *tola* for the same amount of gold collected by the *Tekelas* and the *Baras*, and ¼ *tola* by the *Bhandar Kakati*.[138] Over and above, the officers directly connected with the treasury used to take ¼ *tola* of silver for each *tola* of gold stored.[139]

The amount of gold collected annually to the royal treasury was based on prior estimation. Maniram Dewan reports that from upper Assam alone about six to seven thousand *tolas* of gold were collected during the reign of king Rajeswar Singha.[140] The amount of gold revenue, however, gradually decreased in the period after the civil war, so that during Gaurinath Singha's reign the amount collected was about four thousand *tolas* only and that during the close of our period it was only two thousand *tolas*.[141] During this period not only the number of *paiks* engaged in washing gold was reduced to less than one half,[142] but the rate of tax collected was also reduced from one *tola* to ½ *tola* per head per annum.[143] It appears that the reduction was not to relieve the *paiks* of their burden of taxation but to encourage them to be attached with the profession so that the lucrative source of revenue could be continued. It is however to be noted that the rate of tax on the gold-washers from lower Assam was simultaneously increased. Francis Hamilton reports that the gold-washers at the mine at Pakariguri in Darrang were taxed at 1½ *tola* per head.[144] The possible reason was that the gold from this mine was one of the best qualities[145] and the rates might have been increased in order to compensate the sharp need of money from the gold thus collected.[146]

All other artisans like the blacksmiths (*Kamar*), potters (*Kumar*), Weavers (*Tanti*), carpenters (*Barhoi*), iron smelters (*Lo-saliya*), salt miners (*Lon-puria*) etc. had to pay taxes in terms of their produce or had to work for the king for about 3-4 months in a year in producing such items in which they were proficient.[147]

The royal treasury was enriched through many other taxes and levies imposed upon the peasant *paiks* and traders on various occasions, which were not directly related to land. Another head of revenue income was collection from mines and natural resources. Woods, reeds and bamboos were some of the main products obtained from the forests. *Agar* wood which was in great demand in Bengal, sometimes provided the cause of Assam Mughal conflicts.[148] Besides, peasants were to supply the state with wood of different kinds, dye, honey, fuel-wood, animal and animal products. The *Hatichungi khel* of the Maran tribe supplied elephants to the king and his officers annually. The chronicles mention that two officers of the Hatichungi Marans, Naharkhora Saikia and Ragha Neog were panalised by Kirtichandra Barbarua for supplying inferior quality elephants during the time of Lakshmi Singha.[149] From the salt springs of Sadiya and other mines were also collected some amount of revenue. It is recorded that from the salt mine at Sadiya alone, salt worth about Rs. 40,000 per annum was produced.[150] This was what obtained after allowing 1/3 of the total collection to the Naga and Dalfa tribes under whose jurisdiction the mines were located.[151]

Revenue Derived from the Neighbouring Hill Tribes

To the treasury of the state also came tributes and presents from the subordinate hill tribes of the neighbouring hills. The numerous Naga tribes in the south-eastern region who accepted vassalage of the Assam Rajas from the very beginning used to pay tributes which consisted of horses and *kheh* cloths, bisons, goats, cane and chiefly salt and other products.[152] The Naga chiefs usually once in a year used to visit the Assam capital and offered tributes called *palsewa* and presents.[153] But these tribes frequently defied the overlordship of the Assam king and used to raid the bordering areas so that the rulers of Assam were compelled to introduce a system known as *posa*, whereby the Nagas and the other raiding tribes got some right to collect certain articles from the plains.[154]

Among other tribes chiefly the Daflas, the Akas, the Miris, and the Abars used to pay tributes to the Assam overlords. Their tributes included *jaka* (a kind of bamboo basket), bison, manjit, bow, *jabrang* etc.[155] But their chief item of contribution was usually cotton.[156] Like the Nagas, these tribes also obtained from the Assam rulers same facilities provided through the system of *posa*. The Mikirs, now known as the Karbis, occupying the hilly region in the Dhansiri valley on the south of the Brahmaputra, used to pay articles—chiefly cotton, bamboo mats, barks, *sanchipat* (a kind of bark used for writing), *da* etc. which Butler estimated as amounting annually to rupees three hundred and thirty eight.[157] Tributes from other small tribes of the region like the Lalung, Hajong etc. included honey, buffalo-horn, hide, rhinocheros horn etc.[158]

In order to look after the entire relations with these hill tribes, the kings of Assam appointed persons called *Katakis*[159] who were responsible for regulating the whole affairs. Tributes and presents obtained from the hill tribes, were however nominal. But the Ahom rulers tried to obtain it in order to establish their suzerainty over the tribes and for this purpose they had to make even campaigns against them or had to pacify them by granting privileges.

Taxes Other than Land Revenue

Of the important sources of revenue other than land tax was the tributes paid by the tributary rulers and subordinate tribal chiefs called *Rajas*. They included those of Darrang, Rani, Luki, Beltola, Barduar, Bholagram, Mairapur, Pantan, Bangaon, Bagaduar, Dumariya, Neli, Gobha, Chahari, Dandua, Barepujia, Topakuchia, Khaigharia, Panbari, Sora, Mayang, Dhing, Tetelia, Baghara, Na-duar and Bhurbandha.[160] Their contributions included supply of *paiks* usually in times of war[161] and other annual and occasional tributes, paid usually in terms of goods. They also deposited some amount of cash during the later part of the Ahom rule when money entered the markets of the entire north-eastern region. In times of need, the Ahom rulers, however, could collect tribute from them entirely in cash as can be seen from the proposal to collect money from many of the *Rajas* and chiefs, to defray the expenses of the English East India Company's troops, who came in aid of the Assam king during the civil war period. The list was as follows:[162]

Doomariah (Dimarua)	Rs. 5,000
Kamroop (land revenue)*	Rs. 80,000
Derung (Darrang)	Rs. 50,000
Beltalah	Rs. 2,500
Rannygong (Rani)	Rs. 6,000
Noadewar (Naduar)	Rs. 15,000
Chatgurriah	Rs. 2,000
Chardewar (Chariduar)	Rs. 6,000
Geledary (Ghiladhari)	Rs. 3,000
Mainging Panbary (Mayang & Panbari)	Rs. 400
Dhing	Rs. 2,000
Sadtraja	Rs. 1,000
Panchrajah and Babejie	Rs. 1,000
Nagaon	Rs. 2,000
Korringihe (Kharangi)	Rs. 2,000
Chunderiaah	Rs. 400
Doorea (Duwaria)	Rs. 400
Candahar choky supposed collection	Rs. 20,000
Total amount of collection	2,00,700

The next important source of revenue was undoubtedly the customs duties collected on various occasions and places. Inscriptions of medieval period recording grants of lands make mention of a number of such taxes exacted on the peasantry. In this context, mention may be made of *dan*[163] (sales-tax), *hat*[164] (tax on buying and selling in the market), *ghat* (tax for using rivers for trade), *phat* (duty on marketing places), *katal*[165] (payment made in lieu of personal service), *pad* (fee to be paid by *paik* on getting an assignment), *panchak* (contributions), *beth* (to catch wild elephants from the forest for the state), *begar* (forced labour obtained from *paiks*), *chor* (fine taken from thiefs), *chinla* (fine on *paiks* for adultery), *dhumuchi* (king's claim on the properties of those who died without heir), *marecha* (tax for acquiring land through a marriage

* It did not, however, form a trilantary tribal state.

according to vedic rites), *ghamecha* (tax to be paid to the king for marriage), *danda* (fine for crime), *bandha* (keeping in confinement and compelled to do work), *khut* (tax on movable articles), *jalkar* (tax for fishing) *yavakshar* (tax for production of gunpowder)[166] etc. Kirtichandra Barbarua arranged for the following collection as customs duties from the Raha *hat*,[167] a market place on the Mikir and Kachari border.

TABLE 4.1: Amount Collected as Customs Duties from the Raha *hat*

Source	*Amount Collected*
Ox per head	3 *pons* of cowries
Betel-nut per *kaon*	3 *pons* of cowries
Betel leaf per *mutha* (of 20 *guchis* each)	3 *guchis*
Amount of rice to be collected per 20 *mons*	3 *puras*
Amount of silver to be collected for every 20 rupees	3 *anas*
Amount of cotton to be collected for each 20 bales	1 bale
Every trader had to give as *dan*	17 *pons* of cowries

It appears that from every market a fraction of the goods brought for sale, or cash in cowries and silver rupees were collected as taxes. According to Maniram Dewan customs duties collected from various markets and *chokies* during the decadent period of Ahom rule was 2,56,000 rupees.[168] There were also a large number of *chokies*, i.e., customs houses in border regions. Important of these were those at Sadiya, Sala, Raha, Kapilimukh, Kajalimukh, Batakuchi, Jagi and Kandahar. The *Deodhai Asam Buranji* mentions that Momai Tamuli Barbarua established a large number of *hats* and customs houses all over the state.[169] Francis Hamilton records that the royal treasury received as customs duties Rs. 5,000 per annum from the *choky* at Salaphat alone.[170] Another fixed amount, not known for certain, was collected from Raha[171] and the *Ujir Barua* in charge of the Assam Bhutan border gave to the king some presents, collected from the markets under his jurisdiction.[172] The maximum income that came to the royal treasury was from the Kandahar choky, better known as the Assam choky. It was through this choky that the entire trade with Bengal was carried on. At the flourishing state its annual collection was estimated at Rs. 90,000 per year.[173] As we have mentioned earlier even during the critical years of the Ahom rule towards the close of the 18th century, it was estimated that the customs duties on exports and imports through this choky amounted to Rs. 20,000 per annum.[174] During 1807-08 it was leased to two *Baruas* each of whom used to pay each year to the royal treasury Rs. 45,000.[175]

The list as mentioned above covered a wide range of taxes extracted from the peasant *paiks* and traders. Records from the chronicles show that even the small traders trading in betel-nut and betel-vine had to pay taxes. The chronicle mentions how one Roumuria Bharali Barua imposed taxes on every small trader (*pahar*) at the rate of 4 cowries per month.[176] Taxes at the rate of ¼ of the total produce was imposed on those who had *muga* silk farm and sold fishes.[177]

Forced labour without any sort of remuneration called *begar*, formed one of the most important extra burdens on the peasants. In Assamese society the term *begari khata* in the sense of doing labour without wages is still prevalent. During the medieval period, *paiks* were sometimes requisitioned to do additional work for the state over and above their normal duty of giving personal labour. That happened usually in times of war, in clearing new lands in war footing, in carrying stones from distant lands for building purposes etc. In this way a large number of roads and embankments were built in upper Assam with forced labourers. Gadadhar Singha is said to have built the *Dhodar Ali*, the road running from Jaipur to the east of the Dhansiri river which is still extant. His other such big roads include the *Cheuni Ali* in Sibsagar, now a part of the National Highway No. 37. Both the roads were built with forced labour requisitioned from the *Satras*.[178] It is also recorded that king Udayaditya Singha finding that the *Satras* became full of *bhakats* (devotees) as means of avoiding corvee labour, took-out most of them and forced them to work in the *khats* as *khatowals* and *bebejiyas*.[179] This sort of forced labour, in course of time, resulted in a series of peasants' protests in our period so that the peasants of Nagaon and Darrang came in open protest against the system.[180]

From the above it appears that no able bodied men or any piece of cultivable land was left unused for enrichment of the state treasury during the medieval period. All stress on land survey and measurement were laid and all precautions taken to realise this objective. As the state supplied every *paik* with land for his maintenance and regulated his labour and claimed as due a major share of whatever he produced, the *paik* was deemed as contract labour and in that sense, he was a sort of state slave. Under the circumstances a very small amount of surplus was retained with the peasant *paiks* and artisan classes, so that hardly any trade worth the name, on individual lines, could develop. The only trade that these people did, was the transactions of smaller things necessitated for their daily uses.[181]

Taxes and all sorts of revenues, mostly in form of personal labour and produce, and lack of trade and commerce worth-noting meant an undeveloped and feudal economy.[182] That was exactly the situation during almost the major period of Ahom rule in Assam. It was only after they had come into contact with the Mughals who had a developed money economy, that the Ahoms gradually turned their attention to change their economic pattern.[183] As a result, towards the close of the Ahom rule, monetisation of the economy slowly came in.

Though no surplus remained with the *paiks*, the amount of produce which the *paiks* actually retained with them, over and above the amount paid as taxes, was sufficient to run a family for the year.[184] It was because, every body, except the slaves and bondsmen, had possessed a piece of land sufficient to provide for a family's living. The supply of lands to the able-bodied men was thus a state policy so that the state working force could be strengthened. In addition, the garden or the orchards which every working peasant possessed within the campus of his homestead, was sufficiently big so that it could provide the peasant with almost everything of daily use. The incentive that the peasants got from the state to clear new lands[185] for cultivation of some seasonal crops like mustard, pulses, sugarcane, gourds etc. with no revenue or little revenue, entitled them, at least temporarily, to possess certain amount of additional fertile lands. Thus, even though the Ahom revenue and civil administration possessed many characters of a feudal society, it had some safety valves in the form of liberty and possession so that the peasants here were not like the serfs in a typical feudal social system.

REFERENCES

1. Mackenzie, *op. cit*, p. 6; *Mills Report*, p. 2; *DAR*, p. 199; Gohain, *op. cit*, p. 114.
2. See *Supra*, pp. 53f.
3. The process of Ahomisation involved both inter-marriage with the local tribes and their inclusion into Ahom social hierarchies. Members of many of the local tribes, like the Marans, Barahis, Chutiyas and others became thus Ahoms. Some times, assignment of higher posts to individuals of other tribes led ultimately to include the family of such individuals into the Ahom clans. One of such converted Ahom family was that of Parbatiya Bargohain (also called *Bhandari Gohain*) who actually was a Chutiya. See *ADB*, p. 83.
4. See p. 6.
5. *Ibid.*, p. 98. Also see, *SAB*, p. 3.
6. *Ibid.*, p. 103; also *ibid.*
7. *Ibid.*, p. 16; *HAG*, p. 86.
8. *HAG*, p. 86.
9. *SAB*, p. 21.

10. *PAB*, p. 60; *ADB*, p. 119. The *PAB* mentions that king Pratap Singha, in order to ascertain the number of *paiks*, once personally counted the men by gathering them on the bank of the Sukansuti river (Tuni). See p. 60.
11. For details see *Infra*, p. 185ff.
12. *PAB*, p. 60; *SAB*, 21.
13. *DAB*, p. 130; *ibid*., p. 77; *ADB*, p. 123.
14. *PAB*, p. 60; *ADB*, p. 123.
15. Gohain, *op. cit*, p. 154. For Momai Tamuli's establishment of new villages. See *DAB*, pp. 130ff.
16. *DAB*, p. 130; Gohain, *op. cit*, pp. 154f.
17. *Habib(B)*, pp. 63f; Sarkar, *op. cit*, p. 78.
18. Sarkar, *op. cit*, pp. 78f.
19. Gaunabhiram Barua 'Agar Din Etiyar Din' in *Asam Bandhu (1885-86)*, (ed.), N. Saikia, Guwahati, 1984, p. 75; also *ADB*, p. 499.
20. *PAB*, p. 60; *ABSM*, p. 63; *SAB*, p. 26; *ADB*, p. 500.
21. Both the *DAB* and the *SAB* record that the Maran-Barahis together formed some four thousand souls and the Kacharis twelve thousand. See *DAB*, pp. 100f; *SAB*, pp. 5f.
22. *SAB*, pp. 16ff; *HAG*, pp. 86ff; *CHAB*, pp. 198f.
23. *News Letter of the Court of Emperor Aurangzeb*, dated 10th December, 1669, *Akhbarat-i-Darbari-Muala*, Royal Asiatic Society, Ms. 12-11; cf. Sarkar, *loc. cit*; Bhuyan, *op. cit*, p. 2f.
24. Wade, *op. cit*, p. 132.
25. S.K. Bhuyan has computed the number of adult-male population of Assam to be around 2,40,000 about mid-18th century. See *op. cit*, p. 2.
26. *MEAG*, p. 57.
27. *Asam Bandhu*, p. 78.
28. *MEAG*, p. 57.
29. *Asam Bandhu*, p. 74.
30. *SAB*, p. 77; *PSN*, Bhumika, p. 112; Mackenzie, *op. cit*, p. 6; *Mills Report*, p. 2; *Asam Bandhu*, p. 76; *ADB*, p. 496.
31. *Jenkins' Report in Mills Report*, p. 87; *PSN*, Ins. Nos. 39, 79, 132; *DAR*, p. 201.
32. Francis Hamilton while preparing his *Account of Assam* in 1808/09 writes that the king "grants a part of them (i.e., the *paiks*) to his officers for their maintenance and for support of their dignity so that there is no issue from the treasury for the pay of any officer nor indeed to any person except the foreign soldiers, merchants and mendicants. The officer either accepts of the composition or employs his *Payiks* to cultivate the farms (*khat*) which supply his family with provisions to build his houses to make and man his boats and to make his furniture and clothing; so that his outlay of money is very trifling" *AAH*, p. 23.
33. Vol. I, p. 257.
34. *DAR*, p. 201; David Scott notes that the amount varied from Rs. 6 to 18 per *got*. See Mackenzie, *op. cit*, p. 380. The variation in amount is not unnatural in the sense that artisan class *paiks* had to pay more than the peasant *paiks*. See *Infra*, pp. 169ff, also see *HAG*, p. 240; *ADB*, p. 496; Gohain, *op. cit*, p. 117; *CHAB*, p. 394; Bhuyan, *op. cit*, p. 11.
35. See, *The Comprehensive History of Assam*, II, p. 19; *The Cambridge Economic History*, I, p. 488. The issue of coins regularly by the Ahom kings can be dated since the period of king Siva Singha (1714-1744) only.
36. *DAB*, p. 108. Also see *PAB*, p. 61.

37. The *chamuas* not holding an office, had to pay Rs. 1½ each to the state per annum. See *Asam Buranji* (Ms) by P. Naobaicha Phukan.
38. *Note by David Scott* in Mackenzie, *op. cit*, p. 380; *Asam Bandhu*, p. 74; *CHAB*, p. 394; *MEAG*, p. 86.
39. *Mills Report*, p. 2.
40. This tax, continued also by the Burmese during their rule over Assam (1821-1825 A.D.), was also known as *Kharikatana*. See, Bhuyan, *op. cit*, p. 530.
41. *Ibid.*
42. *Note by David Scott* in Mackenzie, *op. cit*, p. 378.
43. Dhekialphukan, *op. cit*, p. 52; Gohain, *op. cit*, p. 118; *ADB*, p. 496.
44. *Mills Report*, p. 2; Mackenzie, *op. cit*, p. 6; *Buranji Vivek Ratna*, II, (Ms); *ADB*, p. 496; Powell, *Land Systems*, p. 401; Major A. White, *A Memoir of the Late Davit Scott*, Guwahati, 1988, p. 209.
45. *Mills Report*, p. 3; *ADB*, p. 497.
46. Mackenzie, *op. cit*, p. 6.
47. See *Infra*, pp. 243ff.
48. *Mills Report*, p. 2.
49. *HAG*, p. 143; Sarkar, *loc. cit.*
50. Bhuyan, *op. cit*, p. 11; *CHAB*, p. 396.
51. See for details on the revenue of the artisan classes collected in terms of cash and their produce, *Infra*, pp. 169ff.
52. *ADB*, p. 464.
53. Royal manufactories for production of the ornaments, boats, cloths etc. were known as *Rajsal* and the king had a number of *sals* in different places of the state. See *ADB*, pp. 462f; Wade, *op. cit*, Introduction, p. XXXV; *KGC*, p. 251.
54. *Infra*, pp. 146f; also *ADB*, p. 468.
55. *Asam Bandhu*, p. 290.
56. *KGC*, p. 55.
57. ***Kamrupar Buranji***, p. 31; ***Bharistan***, 1, p. 272.
58. *Ibid.*; *PSN*, *Bhumika*, p. 111.
59. A. Guha, *Jamidarkalin Goalpara Jilar Artha Samajik Avastha*, Dhubri, 1985, p. 5.
60. *Ibid.*
61. *Ibid.*
62. M'Cosh mentions that the share of a *Choudhury* was 7 per cent, that of a *Thakuria* was 4 per cent, *Patwari* 1 per cent and *Gaon Kakati* 2.5 per cent of what they collected; *op. cit*, pp. 121f. Besides for every 100 rupees collected, a *Choudhury* got 1 *pura*, a *Thakuria* 6 *Puras*, *Changkakati* 4 *puras* and *Patgiri* ½ *pura* of land. See *op. cit*, p. 122. This shows that the officers were encouraged by the system itself to increase the number of 'hundreds' so that they could get more land. Also see *CP Inscription recording the Settlement of Land in Bausi Pargana, Saka*, 1738 (1816 A.D.), *PSN*, Ins. No. 101.
63. *ABHB*, p. 129.
64. M'Cosh, *op. cit*, p. 121.
65. *Asam Bandhu*, p. 290.
66. *PSN*, Ins. Nos. 73, 75, 76, 90, 91.
67. *Op. cit*, p. 121.
68. *PSN*, Ins. No. 101.
69. *AAH* p. 36; *Kamarupa Purabritta*, cf. *PSN*, *Bhumika*, p. 111.
70. Dhekialphukan, *op. cit*, p. 52; also see *Supra*, p. 146.

71. *Kamarupa Purabritta*, cf. *PSN*, *Bhumika*, p. 111; *KGC*, p. 55.
72. *Mills Report*, p. 2fn.
73. *Ibid.; AAH*, p. 33.
74. *AAH*, p. 33.
75. *Ibid.*, pp. 33ff; Martin, *op. cit*, p. 622.
76. *Rajeswar Singha*, p. 245.
77. *Infra*, p. 247.
78. *AAH*, pp. 30ff; *Bharistan*, I, p. 413; *Kamrupar Buranji*, p. 35; *Asam Bandhu*, pp. 10f; *CHAB*, p. 14.
79. *DAB*, pp. 67, 140.
80. *Ibid.*, p. 139.
81. *AAH*, p. 31.
82. The number of *paiks* varied from 10,000 maximum to 3,000 minimum according to circumstances, e.g., need of the Barphukan or the capacity of the Darrang Raja. See *Asam Bandhu*, p. 10; Bhuyan, *Rajeswar Singha*, p. 142. These *paiks* included those of *Gohain Hazar, Kathar Hazar, Kalita Hazar, Khatar, Hazar, Mahima Hazar* and *Athiyabariya Hazar*, See *CP Inscription of Pahumariya Gosain, Saka* 1738, *PSN*, Ins. No. 152.
83. *Asam Bandhu*, p. 10.
84. *AAH*, p. 48.
85. *Ibid.*
86. Bhuyan, *op. cit*, p. 266.
87. *Ibid.*
88. *Lakshmi Singha Buranji* (Ms); Bhuyan, *Rajeswar Singha*, p. 243.
89. Bhuyan, *op. cit*, p. 329. For the document of the treaty concluded between Assam king Gaurinath Singha and Captain Welsh on 28th Feb. 1793, See C.U. Aitchison, *A Collection of Treaties, Engagements and Sanads . . . etc.*, Vol. 12, Reprint, Delhi, 1983, pp. 112ff; also Bhuyan, *op. cit*, pp. 338ff.
90. The *Saharias* were civil and revenue officers under the Koch rulers of Darrang. *TB* (Eng) p. 238; *ADB*, p. 548.
91. *Buruk* or *Baruk* was a petty officer who was entrusted as a courier, errand man or punishers, *TB* (Eng), p. 229.
92. See Chapter II on classification of lands.
93. *PAB*, p. 61; *SAB*, p. 77; *ABSM*, p. 63.
94. *HAG*, p. 256; Gohain, *op. cit*, p. 116.
95. *Supra*, p. 142fn.
96. *SAB*, p. 77. It appears that Gait while compiling his *A History of Assam* did not come across that fact due to which he says that it was king Gadadhar Singha who first introduced the system of land measurement by using a standard bamboo pole (See p. 171). That the record of the chronicle giving this credit to king Pratap Singha is genuine, may be presumed from his contact with the Mughals for about four decades who had regular land-survey systems.
97. See *Habib(B)*, pp. 377f.
98. *HAG*, pp. 170f.
99. *Ibid.* Also see *Jenkins Report* in *Mills Report*, p. 87.
100. *Ibid.*
101. *CP Land Grant Inscription of Rudra Singha, Saka* 1623 (1701 A.D.); *Nilachala CP Inscription of Madhavadeva* (undated); *CP Land Grant Inscription of Rudra Singha, Saka* 1622 (1700 A.D.); *Dhenukhana CP Inscription of Satyanarayan etc. Saka* 1393 (1471 A.D.), see *PSN*, Ins. No. 15, 108, 110, 106.

102. *Changrung Phukanar Buranji*, p. 15; *SAB*, p. 77, *Kachari Buranji*, p. 93.
103. *CP Land Grant Inscription of the Siva Temple at Dergaon, Saka* 1656 (1734 A.D.); *CP Land Grant Inscription of Sundaridiya Satra, Saka* 1660 (1738 A.D.) etc. See *PSN*, Ins. Nos. 30, 37.
104. D. Chutiya, *Souvenir*, Seventh Annual Conference of the Numismatic Society of India, p. 33; *JHR*, IV, pp. 51f.
105. *KGC*, pp. 141, 311, 506; *Kachari Buranji*, p. 93.
106. See H.C. Barua, *Hemkosh*, 6th edn, Guwahati, 1985, p. 1007. For Hunter's opinion see, *op. cit*, I, p. 47.
107. Hunter, *op. cit*, I, p. 49.
108. *PSN*, Ins. Nos. 89, 98, 99.
109. *Habib(B)*, pp. 377ff.
110. *PSN*, Ins. No. 100.
111. *Ibid.*, Ins. No. 99; also see Ins. Nos. 17, 18, 123, 129.
112. *Ibid.*, Ins. No. 146.
113. *SAB*, p. 77.
114. The *Kachari Buranji*, pp. 93f.
115. *Changrung Phukanar Buranji*, pp. 5, 6, 18, 19, 24.
116. *PSN*, Ins. No. 100.
117. *Op. cit*, Vol. I, p. 133.
118. It is recorded that this system was started from the 19th year of Akbar's reign. See *Ain-i-Akbari*, Vol. II (tr.) Jarrett, Calcutta, 1949, pp. 66; U.N. Dey, *The Mughal Government*, Delhi, 1980, p. 117.
119. p. 93.
120. *PSN*, Ins. No. 97; (See Appendix-E).
121. See for example, the measure of the floor of the Tajmahal in *Habib (B)*, p. 383. For the system in Assam, see Dhekialphukan, *op. cit*, p. 53.
122. The word *darab* means pole or *tar* of measuring lands. So the term *Darabdhara Barua* implied officer in charge of the affairs of measuring and surveying lands. See Rajkumar, *op. cit*, p. 186; *Hemkosh*, p. 492.
123. *HAG*, p. 185; *Jenkins Report*, in *Mills Report*, p. 87.
124. *HAG*, p. 185; *PSN*, Introduction, p. 136.
125. Cf. *PSN, Bhumika*, p. 4.
126. *DRV*, v. 548; Also see *PSN*, Ins. No. 64. From one of the chronicles, it appears that king Suhungmung was the first Ahom king to issue *tamrapatra* for recording the event of land grant in our period. He is said to have issued it to record the assignment of land to the Chutiya prince Sadhaknarayan after the latter's kingdom was annexed by the Ahoms, See *DAB*, p. 200.
127. *ABHB*, p. 124; Rajkumar, *op. cit*, p. 186.
128. For the reissue of the copper plates on the original being lost, brittled or burnt, see the *Nidhanpur Grant of Bhaskaravarmana* in M.M. Sharma, *op. cit*, pp. 38ff; the *CP Land Grant Inscription of Dakshinpat Satra, Saka* 1671 (1749 A.D.) in *PSN*, Ins. No. 52; also see Ins. Nos. 85 and 155.
129. For details see *PSN*, Gȧit has pointed out that king Siva Singha alone issued as many as nineteen such copper plates. See *HAG*, p. 184.
130. See *PSN*, Ins. No. 80, 81. Also see *CP Inscription of Servitors for the Pahumariya Sarujana Gosain, Saka* 1738 (1816 A.D.), Ins. No. 152; (See Appendix-D).
131. *CHAB*, p. 413; *Mills Report*, p. 2; Rajkumar, *op. cit*, p. 216.
132. *AAH*, p. 23.

133. It is found that amount of gold collected for the royal store during Chakradhvaj Singha was five thousand *tolas*, and that during Gaurinath Singha it was four thousand *tolas*. See *ADB*, p. 458. J.N. Phukan mentions that there were two classes of *paik-khels*, one to supply labour, the other to supply articles. See *op. cit.*
134. *ADB*, p. 462.
135. *AAH*, p. 64.
136. *HAG*, p. 143; Sarkar, *loc. cit.*
137. Maniram Dewan, 'Native Account of Washing for Gold in Assam', *JASB*, Vol. VII, Pt. 2, 1838, p. 622; *ADB*, p. 459.
138. *Ibid.*
139. *Ibid.*
140. *Loc. cit*, p. 622.
141. *Ibid.*
142. See the chart in Maniram Dewan, *loc. cit*, p. 621.
143. *Ibid.*, p. 622.
144. *AAH*, p. 47.
145. Wade says that the gold extracted from the river Bargang was of best quality; its price per *tola* being 18-20 rupees. The gold obtained from the river Dikrong was of inferior quality, its price ranged from 12 to 14 rupees per *tola*. See *Geographical Sketch*, p. 20.
146. Revenue collected in terms of gold might have filled up a considerable gap in revenue of the state. For, as recorded in *Fathiya-i-Ibriya* the price of gold in the middle of the 17th century was 8-9 rupees per *tola* (cf. *HAG*, p. 143). The rate gradually increased to 12-20 rupees in the last part of 18th century. See Wade, *Geographical Sketch*, p. 20.
147. *AAH*, p. 146. That the artisan class *paiks* were to work for the king for six months in a year (*ADB*, p. 464) is not tenable as a *paik*, as a rule, was to work for a maximum of four and a minimum of three months in a year.
148. The Mughals before the battle of Saraighat, 1671, demanded from the Ahoms—*agar* wood, elephant, black-peper and gold dust as a venture for any peace proposal between the two (*SAB*, p. 85). Some Mughal traders even secretly entered the forest of Assam for collection of *agar* wood. See *PAB*, p. 63; *DAB*, p. 109.
149. *ADB*, p. 256; Bhuyan, *op. cit*, p. 207.
150. Cf. Pemberton, *Report on the Eastern Frontier of British India*, 2nd Print, Guwahati, 1966, p. 84; *ADB*, p. 471.
151. *ADB*, p. 471.
152. *DAB*, p. 35; *ABS*, p. 27; *ABHB*, pp. 40f.
153. *ABS*, p. 19; *ABHB*, pp. 22, 41.
154. *Ibid.*, p. 23; *Ibid.*, pp. 33, 41. Different privileges given to these tribes by the Assam kings include land for rice cultivation called *khat* (like *Naga-khat*) in the foothill areas, fisheries called *bil* or man called *Bahatiya* to supply them with needed articles such as rice, cloth, cattle, or the privilege of having commercial transactions with the plains. See *CHAB*, p. 370; Gohain, *op. cit*, pp. 137f; Devi, *op. cit*, pp. 270f.
155. *ABS*, p. 27; *ABHB*, p. 40; *PAB*, p. 125.
156. *Mills Report*, p. 12.
157. *Travels*, pp. 127f; also Gohain, *op. cit*, p. 139.
158. *DAB*. pp. 134f.

159. *ABS*, p. 27; *ABHB*, pp. 40f; Gohain, *op. cit*, p. 137.
160. Popularly known as the *Satraja* and *Panchraja*, those tribal chiefs included all those excluding the chiefs or *Rajas* of Darrang, Rani and Beltola. See *DAB*, p. 143; Bhuyan, *Rajeswar Singha*, p. 143.
161. Rudra Singha while preparing to invade Bengal is stated to have collected 40,000 *paik* soldiers from the tributary *Rajas*. See Wade, *op. cit*, p. 132. The *DAB* (p. 143) mentions that the *Panchrajas* supplied king Rajeswar Singha with 40 people. Also see, *Kachari Buranji*, p. 99.
162. See the Treaty of commerce concluded between Captain Welsh and king Gaurinath Singha, dated 28th Feb. 1793 in Aitshison, *op. cit*, pp. 112f; also in N.N. Acharya, (Comp) *Historical Documents of Assam and Neighbouring States*, New Delhi, 1983, pp. 11ff; Bhuyan, *Rajesewar Singha*, p. 246.
163. For example, in Darrang people had to pay a tax on every animal sold. See *Supra*, p. 160.
164. There were a number of *hats* or marketplaces established by the Ahom rulers. King Pratap Singha is said to have established *hats* like Namchang, Barhat etc. on the Assam Kachari borders (see *ABHB*, p. 40) Kirtichandra Barbarua established *hat* at Raha (See *DAB*, pp. 141f). A *hatkhowa* was appointed to collect *hat* taxes.
165. See, *PSN*, *Bhumika*, p. 6, and also Ins. Nos. 79, 80 etc.
166. Cf. *PSN*, *Bhumika*, pp. 6f; Ins. No. 64; also see *SAB*, p. 29. Towards the close of the Ahom rule the rulers introduced some new taxes, e.g., *khatiyadhan* or a kind of poll tax and *dewaldhan* or a land tax. See, Maniram Dewan's reports to Major A. White (a) Self-Dependent Assam, Cf. B. Sharma, *Maniram Dewan*, Ist edn, Guwahati, 1950, Preface, pp. IVff.
167. *DAB*, p. 142.
168. Cf. *Mills Report*, Appendix-KB, p. 621.
169. *DAB*, pp. 130, 142.
170. *AAH*, pp. 47f.
171. *Ibid.*, p. 48.
172. *Ibid.*
173. *Bengal Political Consultation (Welsh to Cornowallis)*, March 11, 1793, No. 17.
174. *Supra*, p. 178.
175. *AAH*, p. 42.
176. *SAB*, p. 29; *ABS*, p. 27; *ABHB*, p. 40.
177. *SAB*, p. 29.
178. Hunter, *op. cit*, I, p. 258; Gohain, *op. cit*, p. 155.
179. *KGC*, p. 560. The *Bebejiyas* were the adult-males reserved for inclusion (*khajithowa*) as *paik* into any *khels* when some members of that *khel* died or absconded. Usually those *paiks* who left their own *khels* and absconded, or those who were awaiting registration as *paik* were reserved in this category. See Bhuyan, *Rajeswar Singha*, p. 247.
180. *TB*, pp. 119f; Wade, *op. cit*, pp. 243f; *CHAB*, pp. 323, 325f.
181. *Infra*, p. 237.
182. R.S. Sharma, *op. cit*, p. 1.
183. The contact brought to the Ahom court many other changes in form of food habits, dresses, pursuits in fine arts including music and musical instruments and in many other aspects.
184. Also see R.S. Sharma, *op. cit*, p. 1.
185. See *Supra*, p. 41.

5

THE PEASANTRY
Their Material Conditions

The state depended for its growth and existence on the peasants, artisans and workers including slaves and servants who formed the body of the labour force. The aristocracy and the nobility for their maintenance and the peasantry for their own living had to depend on the working classes. Indeed, it were the peasantry who provided all that was required for the growth and development of the state. It was therefore not the soil alone but also the dwellers thereon who were regarded as property of the state.[1] The Ahom rulers of medieval period therefore organized the peasantry in such a way that the state could obtain the maximum use of its labour force. This systematic organisation to which we made a reference earlier, was the *paik* system, i.e., the system of organisation of the labour force of the state.

(A) THE PAIKS: THEIR ORGANISATION AND WORKING

The state population under the Ahoms could be divided into three broad categories—the first two classes were the aristocracy and the gentry, and the third consisting of the vast commonality, i.e., the *paiks*, slaves and servants was the actual working force of the state. The aristocracy included the king, his family members, his kiths and kins and all administrative dignitaries both Ahom and non-Ahom, and all ecclesiastical heads. The gentry consisted of the officers, to the post of which members of the Ahom nobility were not normally recruited.[2] They enjoyed certain privileges in the form of lands and man power—*paiks*, slaves and servants; and received presents from the peasants and the officers junior to them. This class belonged to the non-labouring class of peasants known as *chamua* or *apaikan chamua* who obtained exemption from manual labour for holding a post or by paying a poll tax.[3] The status of a *chamua* was given to those only who belonged normally to a good caste or birth.[4] Thus caste or political background determined the status of the individuals in the society. The Brahmanas and the descendants of the higher officers got by birth the status of a *chamua*. The regular priestly class were exempted not only from physical labour but also from paying poll taxes. The Brahmana priests were to perform *puja* or worship at the houses of the king or the state dignitaries,[5] and the Ahom priests, in a like way, did all customary socio-religious functions for the welfare of the king and his people.[6] Chronicles and epigraphic records of our period show that when satisfied with the service of a *paik* having talent or merit, the king promoted him to the *chamua* rank. Thus king Pratap Singha being impressed by the performance of Momai Tamuli as a gardener who was then rendering service to his master as a bondsman, released him from bondage and directly made him a *chamua* by appointing him as the royal *Barichowa Bartamuli* or the head of the garden-keepers. Thereafter, Momai Tamuli served in different capacities of *chamua* offices till finally he was promoted to the post of the *Barbarua*.[7] A copper plate inscription of king Gaurinath Singha also records that the king being satisfied with the work of one Balaram, a *kanri paik*, appointed him as *Barkataky* (head of the envoys) of Guwahati.[8]

(II)

The actual labour force was formed of the peasant population known as the *kanri paiks* or simply *paiks*[9] who were to produce almost all the necessaries of life and items of comfort. The word *kanri* means those who fought with bows and arrows[10]—indicating a primitive stage of warfare with which they were associated. During the period under discussion when agriculture was the main source of subsistence and keeping a regular and strong standing armed force was impossible, every individual and capable *paik* was to engage himself in agriculture as peasant and warfare as soldier on behalf of the king. This was how the growth and security—the two major aspects of the state depended largely upon the peasant *paiks*. At the same time the nature of the economy of the state also made it pertinent for the rulers to use the man-power to the maximum for production purposes. Economy of the time being still mainly backward when use of money was limited and trade among the peasantry being largely absent, everything was to be produced at home. The Ahom rulers therefore, not only categorised the *paiks* to suit the production purposes but also imported a number of skilled ones for specific professions even from outside Assam.[11] The advantage for collection and organisation of the peasants into a strong working class came from the gradual expansion of the state. As we have shown elsewhere by the end of the 16th century both lands and labours provided the scope for their proper and productive uses.

The extension in territory and increase in the population had certain far-reaching consequences in the organisation of the peasant *paiks*. The expansion of the state confronted the Ahoms not only with the neighbouring states but also with the mighty Mughal power in the west behind whose north-eastern policy were the ideas of territorial expansion dictated by economic and strategic reasons, and proselytization of the conquered infidals.[12] Besides, the natural wealth of Assam, particularly aloe wood and elephants, were very much esteemed in Mughal India. Such circumstances compelled the Ahoms to face a number of invasions from the west during almost the entire 17th century and in the absence of a strong standing army they had to organise the peasant militia in the best possible way, whereby every adult and capable peasant between 16-50 years of age was to be enrolled as peasant soldier. As a part of this organisation, came the need of pursuit of different professions like making of gun-powder and all other war materials. The circumstances demanded, at the same time, the need of economic development through the best utilisation of the soil. The Assamese chronicles make mention of two Ahom monarchs, one of whom fought for about four decades against the Mughals and the other prepared the state for the famous battle of Saraighat (1669-71) when Ram Singha, the Raja of Amber, and the general of Aurangzeb, was defeated.[13] The first one was king Pratap Singha who, as we have mentioned earlier, began organising the *paiks* into a well-knit system, and the second was king Chakradhvaj Singha who prepared his men to successfully withstand the Mughal attack.[14] The second impetus for the organisation of the *paik* system came from the increase of the population which provided the best opportunity for deriving the optimum production from the soil. But the character of the population complex brought with it certain limitations to the field of production. Besides Hinduised and non-Hindu tribes, the population of the time contained certain number of acculturated groups who in the Hindu social hierarchy claimed higher status. The latter included the Brahmanas, the Kayasthas and the Kalitas,[15] some of whom particularly the Brahmanas, as a matter of high ritual status, did not take to plough or other forms of hard physical labour associated with production.[16] The

Ahom rulers, who had in the meantime, became acquainted with the Hindu way of social structure based on castes of different socio-ritual status, and also with the Brahmanical religious practices, recognised the cultural superiority of those people who were at the top of the Hindu social hierarchy and gave them the privileged status of the *chamua*. This had been reflected in the organisation of the *paik* peasantry so that in the ultimate end the position of the *paiks* in the system also became largely based on caste structure. This fact would normally lead to the conclusion that in medieval social system of Assam caste and class were inseparably related.

(III)

The essence of the *paik* system was to requisition through an organised form the entire labour force of the state and to use it for any kind of state service. Assamese chronicles mention how the Ahom rulers of medieval times engaged from the beginning their population in opening jungle lands and bringing them under regular cultivation. The system begins with estimation of the population and its division into castes and professions. After estimating the numerical strength the entire population was divided into two broad categories—one to be utilised for giving personal manual service to the state and the other for giving non-manual service. The number of the latter was smaller in comparison to that of the first and their form of service to the state was considered dignified.[17] It was this class which formed the intermediate group between the aristocracy and the nobility at the top and the peasants or *kanri paiks* and slaves and servants at the bottom.

The permission of paying poll tax instead of giving physical labour was also extended to some of the artisan classes[18] who, however, contributed a share of their produce to the state. This, therefore, in essence, did not give them the *chamua* status whose privilege it was to go away with manual work. Those tax paying *paiks* whose birth was not considered as good or noble, were classed as *paikan chamua*. The rest of the working people, termed as *paiks* (or *kanri paiks*), were to render personal service to the state in stead of paying any money tax. Although theoretically, they too were eligible to have the option of paying money tax in lieu of physical labour in certain circumstances,[19] this privilege was only an exception, not a rule. The *kanri paiks* were thus the actual source of state revenue and were therefore carefully organised.

The organisation of the *paik* population was very interesting and accurate. The smallest unit of the *paiks* was known as *got* in which, in the earlier part of its organisation, were four individual *paiks*,[20] and each one of them had to serve the state for three-months in a year. But the numerical strength of the *got* was reduced to three in the later part of the Ahom rule due to shortage of required man power.[21] Consequently the king had to maintain the balance of the state labour force by reducing the strength of the *gots* of the *paiks* from four to three. Each of the *paiks* in every *got* was to serve the king or the officers of the state to whom he was assigned on rotatory basis.[22] During this period the other members of the *got* not only had to supply him with his fooding but also had to work in his field.[23] In times of emergency—such as war, this proportion of $^1/_4$ or $^1/_3$ of the *got* was raised to $^1/_2$ or even $^3/_4$.[24] In usual times *paiks* were, besides agriculture, engaged in constructive public works. As Gait points out: "In times of peace it was the custom to employ the paiks on public works; and this is how the enormous tanks and the high embanked roads of upper Assam, which are still a source of wonder to all who see them",[25] were built.

The state control over the *paik* was very rigid. They were organised into some bigger units known as *khels*.[26] Each *khel* of *paiks* consisted of 1000-6000 men and was placed under state officers of different grades called *kheldars*[27] like *Phukans, Baruas, Rajkhowas*.[28] Usually an important *khel* like the one engaged in boat building (*Naosaliya*) or manufacturing of gunpowder (*Khaigharia*) was placed under a *Phukan*, and smaller ones were put under a *Barua* or a *Rajkhowa*.[29] In every *khel* again, there were a number of subordinate officers known as *Bara, Saikia* and *Hazarika* under its *kheldar* each supervising twenty, one hundred and one thousand *paiks* respectively,[30] who helped their superior officers in all organisational matters. Every *Bara* in the lowest level of the organisation knew his *paiks* and could mobilise them for any service of the state within a very short time.

The area-wise organisation of the *paiks* in the initial stage created some problems in obtaining from them speedy and uniform service, for a certain area might contain *paiks* of different professions. Under the circumstances, *paiks* were later on reorganised into professional *khels* like arrow-making, bow-making, boat-building, catching and training of elephants, iron-smelting, gold-washing, silk-thread spinning and cloth-weaving, gun-powder manufacturing etc. with members of the same or allied professions, which needed their close settlement in a particular locality. Thus in course of time, *paiks* of a particular *khel* were established either in a single village or in a specified locality. This subsequently brought a peculiar character to the villages of Assam in our period in the sense that a village contained *paiks* of the same caste, for artisans and professional classes were members of some specific caste groups.[31]

(B) THE PAIKS: THEIR POSITION IN THE SOCIETY

Every *paik* for his service to the state received a piece of cultivable land, and thus the *paik* system negated the number of working man who possessed no land in our period; and this fact provided every peasant almost with the minimum need for his subsistence. It is to be noted that although like a bondsman a *paik* had to render his compulsory manual service and other kinds of duties to the state, he was awarded the right to nominate his immediate officers like the *Baras* and *Saikias*, or sometimes even the *Hazarikas* and could demand their dismissal.[32] Towards the later part of the Ahom rule, this right was extended to cover the dismissal even of the great *Gohains*. This shows that although in regard to service a *paik* could be considered as a bondsman, in regard to possession of certain political rights he was sometimes treated as a free citizen. Gait states that this was a most valuable privilege whereby the *paiks* were saved from much of the oppressions that would otherwise have fallen to their lot.[33] In spite of all this, there was a tendency among the officers to take bribe from the *paiks*, in the form of labour service,[34] or produce, so that even roots, fruits and vegetables collected by them were not spared.[35] But when detected or complained against by the peasants, the king dismissed the accused officer. Towards the later part of the Ahom rule, bribery and extortion of some officers reached such an extent that the oppressed people were compelled to lodge complaints before the king even against the nobles and the great officers; and it was then that they were given the right to demand the removal of the higher dignitaries of the state as well.[36]

The *paik* system was not, however, pleasing to the peasantry who were already becoming dissatisfied with the working of the system specially when the duration of their service to the state was increased from three to four months in a year. Besides, they were now made to pay poll taxes on account of their holding a homestead which was earlier rent free.[37] Even the excess lands which the peasants had cleared out from jungles and riverine tracts or in the hill slopes as we have earlier discussed, were becoming subject to taxation and were reclaimed to fill up the gap in the regular quota of lands after every new survey.[38]

The *paiks* were handicapped in their material life in many other ways. Although they could form *khels* of certain professions, they could not form commercial guilds of any sort. The way they were to pay for the state, left for them very little or no surplus. Besides, being "tied to fixed *khel* or territory, they were not free to pursue independent trade and thereby to improve their material conditions. Their way of life was fixed by the state, which they had no right to alter".[39] Indeed the entire *paik* structure point to a state of serfdom bound to the king.

In lower Assam and Darrang where the peasants were not organised into *khels* there they were more oppressed by the revenue-cum-administrative officers of different higher and lower grades. The system of collection of revenue in cash Kamrup created a demand for money in the villages which finally led to debt slavery in the society. The Darrangi peasants besides paying all kinds of taxes, were also to supply working *paiks* for the service of the state who had to maintain themselves at Guwahati with what little amount of food they could carry with them from Darrang.[40]

Like any other consumption goods, the *paiks* formed a part of the salaries of the Government officers and also the means of maintenance of the state retainers. Such *paiks* were known as *Likchous*[41] who were not required to go to the battle field as soldiers. Their status was like that of a semi-slave; they worked for their master, but maintained themselves from their own earning procured from the two *puras* of *rupit* lands allotted to each of them as *paik*. There was another class of *paik* called *Dewaliya paik* attached to the numerous religious institutions like the *Satras* and the temples. These *paiks* were free from performing compulsory military duties for the state. It is to be noted that while *paiks* of such cadre worked for the institution or the person attached to which they were attached, they also like the *likchous*, had to maintain themselves from their own resources. That the lot of the peasant *paiks* was no better than any kind of bonded man can be presumed from the fact that some *paiks* in the border areas near the hills were put under the mercy of the raiding hill tribes who were permitted to collect certain amount of their (the *paiks*) produce so that the rage of the inroading hill tribes could be mitigated.[42] These *paiks* in the status of the slaves, were therefore, known as *Bahatiyas*[43] who had to carry on the commands of both the tribe masters and the state officers. For, these *paiks* were to bear at the same time the obligation to the state also. Thus the overall burden of taxes and imposts on the *Bahatiya paiks* was heavier than that of those *paiks* who were in the main land of the state. Under such conditions these *paiks* may also be said to have been in semi-slavery status. Their only difference from the slaves was that they had a piece of homestead and they could not be sold in the market.

The administration being based on gradation in the official rank from the king at the top and the peasantry, slaves and servants at the bottom, strict distinctions between grades in all sorts of consumption and social behaviour were to be maintained.[44] This was why in spite of being the main source of production, the *paik* peasants were deprived of the right to consume or use their production as they desired or needed. For example, the silk-thread-spinners and weavers or the gold-washers and goldsmiths had the obligation to produce but not the right to store or use any amount of their products in surplus form.[45] Even the lands which the *paiks* obtained as means of subsistence, in fact, did not belong to them on personal basis. It was through the *khels* that a *paik* got his allotment of lands on communal basis and that too he could enjoy during the period of his service only. But at the same time, both the aristocracy, and the spiritual heads enjoyed numberless acres of lands opened for them by their *paiks*, slaves and servants.[46] In this context it should be mentioned that while a *paik* lost his share of lands when his service was no longer required by the state, the state-dignitaries could retain their share obtained from the state as *manmati* or lands received for holding a socio-political status.[47] Thus so long the *paiks* were useful to the state, they were fullfledged land holding peasants, otherwise they formed a wretched class and were at the complete mercy of the state.

(II)

The organisation of the *paiks* on the basis of caste and good birth, led to the growth of an economically and socially privileged class in the society distinguished by caste and political status excluding the higher military services to which the noble Ahoms were normally recruited, the other non-military posts or high civil offices like those of *Phukan, Kataki, Barua* and *Kakati* were filled up from among the Brahmanas, the Kayasthas and the Kalitas. The religious heads were entirely from among the Brahmanas, the Kayasthas and the Kalitas, and some of them were as powerful and wealthy as the king himself and at times dared even to challenge his person.[48] There are ample grounds, therefore, to assume that in the society in medieval Assam, higher the caste, higher was the economy and proprietary status of an individual.

In an agrarian and feudal society, as in medieval Assam, lands and labourers were the major items of wealth. As a natural consequence, the proprietors used the manpower like a commodity over which they had all power and control. The *paik* system of exacting forced labour resembled in some sense an organised form of slavery in the society.[49]

But the system contained certain elements whereby even within its steel frame the *paiks* enjoyed certain rights and privileges and freedom of work, making it different from a slave society. Thus a particular peasant could retain for his own, $^3/_4$ or $^2/_3$ part of his labour, and unlike the slave, he had a plot of land for his own maintenance, and the permission to acquire extra lands by clearing jungles. He had also the option at least theoretically, to be exempted from personal labour if he could pay for it. Besides, as we have mentioned earlier, a *paik* had his say in selecting or dismissing his immediate officers. Sometimes, they were also promoted to the *chamua* status because of their proficiency in work. All this made the *paik* society of medieval Assam different from a slave society. Indeed, the *paiks* were free peasants even though they were bound to serve the state without protest.

Exaction of forced labour at the will of the state against the will of the labourer, gift of peasants along with lands, exercise of powers over them by a number of officers of different grades and the state-policy to apportion their labour and produce to meet the salary of the officers, and more notably, the compulsory service requisitioned form the armed forces, made the system a feudal one.[50] And once it became feudal, peasants were much oppressed both emotionally and physically so that the *paik* organisation gradually became unpopular among the *paiks* who preferred either to escape from the system or protest against it. According to A. Guha, the 'conflict on the question of both the for and the quantum of rent payments between the ruling class and the subjugated people consisting of peasants, artisans and traders emerged as a major condition in eighteenth century Assamese society.[51]

Even in the initial stage, the people protested against their employment to do compulsory manual service.[52] To the *paiks* both compulsory military service and manual labours were irksome. But as because money was neither available nor it was possible for them to pay revenue in cash, the *paiks* silently submitted to the system and were looking for any opportunity to escape from it. It is found that the *paiks* in order to avoid that compulsory manual service, became *bhakats* or residential devotees of the *Satras*, who as a matter of law, were permitted exemption from such compulsion.[53] Mills and Robinson—both noted the fact that a good number of *paiks* accepted slavery as better means of living than being a manual *paik* worker.[54] An interesting practice of avoiding the acceptance of the liabilities of the system was that boys even of fifteen or sixteen years of age which was taken as the age of entrance for a *paik* career, were trying to avoid wearing cloths as far as possible to show that they were still minors, lest they would be considered as adult fit to have their name included in the *paik*-list.[55]

Such escapism, however, did not serve the *paiks* for a better lot in the long run. As a result they were to resort to public protest against this semi-slave and semi-serf system. Their first and the most successful and organised protest took the form of the great Moamariya rebellion,[56] which occurred during the reign of king Lakshmi Singha (1769-80) and continued even after. This rebellion originally started by a section of the *paiks*—the Marans, became gradually spread almost over the entire state and took the form of a civil war. Though apparently the movement achieved nothing more than the mere establishment of a separate kingdom under the suzerainty of the rebels, it produced some far reaching but indirect consequences in the long run.[57] It was for this rebellion that the Ahom ruling class towards the end of their rule, realised the drawbacks of the *paik* system and took steps to gradually abolish it.[58] In the same way some four thousand *paiks* from Darrang, encouraged by the Moamariya rebellion, protested against heavy burden of taxation and compelled king Lakshmi Singha to concede to their demand[59] of suspending the re-settlement of Darrang aimed at increased rate of taxation. Even the social inequality kept between different gradations was to be gradually narrowed down. The Premier Purnananda Buragohain had to accept as *fait accompli* the equality of man by permitting, although in theory, all members of the subject population to use *dola* or palanquine on certain conditions and obtain such other privileges.[60]

Thus by the last part of the 18th century, the social inequality and oppression based on the *paik* system lost its rigidity and by the end of that century these practices became almost obsolete.

The Slaves and Servants

The *paiks* alone did not form the entire labour force in the society. Along with them, there were the slaves and servants called *Das, Dasi Golam* or *Bandi, Beti* (male and female slaves) for whom the state had no obligation whatsoever. Slavery as a recognised institution might have existed in the kingdom of Pragjyotisha-Kamarupa since ancient times.[61] In the period under discussion, the ruling aristocracy and other nobles and tribal chiefs—who formed the top of the feudal gradation in the society possessed large farms cultivated by the slaves and servants. Not only the higher nobles, but men of "respectable position had one or more of them, by whom all the drudgery of the household and the labour of the fields were performed".[62] Keeping of slaves and servants was advantageous in two ways (a) they were exempted from compulsory manual service to the state, and (b) they were to pay no revenue in any other form to the state so that they formed a group of constant labouring class.[63] Slavery thus was a popular phenomenon in the society and it became a recognised institution in the society of medieval Assam; it derived its encouragement from the retaining class itself who were economically much benefitted.

Most of the slaves were of local origin and were products of the economic oppression of the well-to-do class particularly in Kamrup and imbalance distribution of lands between the actual tillers of the soil and the managerial class. The permission granted to the *paiks* to open and cultivate additional lands became meaningless, when during every new survey and settlement, these were categorised as regular lands and distributed among the increasing *paik* population. Even the regularly allotted *rupit* lands were redistributed from time to time so that those, who had disputes with their *kheldars* were compelled to give away their better quality lands and had to be content with some inferior lands.[64] It was but natural that under such conditions those *paiks* who had no source left to maintain their family and to pay taxes took to slavery at the estates of the nobles.[65]

Selling of family members formed one of the sources of slavery. In an old document, dated *Saka* 1607 (1685 A.D.), it is recorded that one Jahari, a *paik*, sold his wife for Rs. 8 to one Gangaram Thakur of Kamakhya of Guwahati.[66] Again one Mahanta of Salaguri Satra was recorded as buying a man for Rs. 6 in *Saka* 1721 (1799 A.D.) sold by twenty villagers.[67] During the Moamariya insurrection and the Burmese occupation of Assam, social condition became so deplorable that many poor families sold their children to make their both ends meet.[68] Thus selling of children by parents due to poverty formed another source of slavery in Assam in the early part of the 19th century when the British had occupied the land. Sometimes poor peasants finding themselves in distress, used to mortgage their own persons to the rich people. This happened particularly in Kamrup where compulsory manual service was computed for money tax and the revenue officers both big and small, harassed the peasants in different ways by compelling them to pay exorbitant amount so as to meet both the actual amount of revenue and the officer's demand of additional benefit. The result was that the wretched fellow finding no way out to gather the amount either fled to the hills or borrowed money from the rich peasants and became a slave, then called *bandha* or bondsman and remained as such until he could repay the debt. The *Katha Guru Charit* mention that one Baloram Atai of Tapa, a village in Kamrup, had mortgaged himself for rupees five and became a slave.[69] Momai Tamuli Barbarua was also a bondsman for Rs. 4 before he was recruited to the king's service.[70] "Such of the peasants", says M'Cosh "have been enthratled by mortgaging their bodies for a few rupees; and for the want of means of accumulating the original sum, increased by exorbitant usury, continue in bondage for life themselves and their discendants from generation to generation.[71] Thus debt-slavery developed as a recognised social phenomenon.

Gifted *paiks* to the spiritual heads or priests, by virtue of such out right gift, became the personal property of their masters. In the early medieval period king Vaidyadeva of Kamarupa (c. 1138-1150 A.D.) through his *Kamauli Grant* bestowed four hundred workmen to a Brahmana named Srutidhar along with lands.[72] During our period the Ahom rulers made huge amount of land-grants to the Brahmanas and religious institutions and bestowed them with gift of *paiks*.[73] One lady of Ahom aristocracy once bought a certain number of slaves and gifted them to the Hayagriva Madhava Temple at Hajo in Kamrup.[74] Thus a servile class resulting out of land grants, grew in Assam in the period of our discussion. Besides, there were presentation of slaves both male and female, as token of cementing friendly relations by the rulers of the neighbouring states.[75]

Criminals, subject to capital punishment were sometimes spared and made slaves for life.[76] In other circumstances, they were bartared as slaves for horses of Bhutan in the foothill markets in the north.[77] Sometimes slaves were also bought from the hill tribes.[78] Most of the slaves, however, were prisoners of war.[79] Besides persons born of slave parents[80] were also considered as slaves and they were not given any right of a freeman.

Once slave, a person became free from all sorts of obligations to the state except for the penalty to be given for offences committed by him.[81] But he/she was at the complete mercy of his/her master and therefore, could be treated like property and bought and sold in the market.[82] The price of a slave varied according to caste and sex. For example, the price of a slave of the Kalita caste, the social status of which was considered as higher, was double than that of a Koch slave of the same age whose caste status was considered inferior to that of a Kalita; and the price of a female slave was half the price of a male.[83] Walter Hamilton reports that girls were chiefly bought for brothels costing from Rs. 12 to Rs. 15 per head.[84] The price ranged from about twenty rupees for an adult male of good caste to three rupees for a women slave of low caste.[85] The mention of prices of slaves of different castes in contemporary records points to the fact that slave trade was an accepted form of transactions and it developed in a larger scale in the period when money gradually entered the markets of the north-east. There are a number of references to slave trade between Assam and Bengal and other neighbouring countries. Francis Hamilton and others mention that about one hundred slaves of pure castes were annually exported to Bengal.[86] According to M'Cosh, although slave trade had ceased to exist by his time (1837) it was once a flourishing trade with Tibet.[87] The districtwise mention of prices of the slaves by early British writers[88] show that slaves were also sold in the local markets of the kingdom.

The extent of slavery in Assam during the period of this study was considerable as it is learnt from different contemporary and semi-contemporary sources. Both Francis Hamilton and Walter Hamilton record that slaves were numerous in Assam, every man of rank having several.[89] David Scott is said to have released twelve thousand slaves at a time in Kamrup alone.[90] From a census taken about the thirties of the 19th century, it is found that of the total population of 2,71,944 in Kamrup, there were some 16,364 slaves and 7,993 bondsmen.[91] It may however be presumable that the number of slaves and bondsmen in Kamrup may be slightly higher than that of other districts in Assam. For, the peasantry in Kamrup, "had been in the habit of paying pecuniary compensation in lieu of personal service"[92] and the exaction was so numerous owing specially to the profit making motives of the revenue officers that bondage and slavery became the way of life for the poor. Besides, the event of political crisis and civil wars towards the close of the 18th century also led to the increase of the number of slaves in the state. As such the number of slaves and bondsmen in some earlier periods might have been smaller. But then there were other servile classes like the *likchous, Bahatiyas* and the *Dewaliya-paiks*. They altogether formed a significant percent of slaves and bondsmen in Assam in our period which is estimated as something around 9 per cent.[93]

Position of the Slaves and the Servants in the Society

The fact that sometimes *paik* peasants preferred slavery to being a free peasant, implied that the life of the slaves in Assam was comparatively less miserable and oppressive. They were mostly used for cultivating the lands of their masters or were engaged by the latter to work for other masters for earning some extra income.[94] For the slaves there was no problem of food and clothing which were supplied to them by their masters. The bondsmen could buy their liberty by repaying their debt accrued along with its interests.[95] Slaves could marry also; but children born of them were normally considered as their masters' property.[96]

Apparently the slaves were almost at par in treatment with peasant *paiks*; and as reported by Jenkins only on very few occasions they were found to have complained against their masters.[97] According to A. Guha, the slaves having "no obligation to fight, its members had a secured life as well. On the other hand free peasantry faced increasing insecurity on account of frequent wars and ruthless exploitation by the official aristocracy which was allowed to exact unpaid *likchou* service from them".[98] But there are evidences to show that except for compulsory military service and other manual works to be rendered to the state, the *paiks* should have no choice for slavery. It can be presumed that it was not because their minds were so prepared so as to silently bear all kinds of hardships as destined by fate. William Robinson's observation in this connection is very much relevant: "these poor creatures feel that they were the personal property of their masters, who have the power to do, as the like with them and it is their duty to submit to all. It is also their impression that any appeal to a public authority will not redeem them from threldom while on the contrary this only will bring upon them the displeasure of their masters and their connections".[99] The biographical works of medieval period on the other hand, contain some examples of maltreatment to the slaves. Thus one slave is said to have complained to a Vaishanva saint that he was neither allowed to sleep at night, nor he was given enough food to eat nor provided with warm cloth; and in spite of his hard work from morning till night, he was frequently beaten.[100] But such examples are very rarely found and can be said to have been applicable in some isolated cases only. On the other hand there were many such examples of good treatment to the slaves in one of which one Chandari Ai was said to have been treated as no less than the mother of the family.[101] In general, therefore, slaves were treated as members of the family.

Eventually slavery had its demoralising effects in the society. Being personal property of their masters, slaves had no property of their own, and whatever they produced throughout their life, contributed to the wealth of their masters and thus slavery created a rich peasant class in the society. Sometimes slaves were engaged by their masters even in committing some crimes like robbery.[102] But this had happened only during the period of the Burmese invasions when plunder and raid were the order of the day as means of collection of wealth.[103] This, however, implies that slaves were sometimes used by their masters to satisfy their greeds or whims or fulfill political interests if the masters belonged to the class of nobility.

Slavery as an institution, however, did not form the basic character of Assam's agriculture in medieval period. It were the *paiks* who formed the major force of production and the entire militiamen. Slaves were not only insufficient to supplement the strength of the peasant *paiks*, but they were also lacking in different specified traits to produce the entire need of the state.

Material Condition of the Peasantry

Not much is systematically recorded in contemporary sources about the material condition of the peasantry. But the items of food, clothing and habitat as referred to in these records would throw some light on this aspect. Rice was the staple food of all sections of the people. But as stated earlier, there were variations in quality of rice consumed by different classes of people and as such the *paiks* could use only the coarse variety except on very few occasions like *Bihu* (national festival of Assam) when they had some finer variety in form of *jalpan* (tiffin). They also used a kind of soft rice called *komal-chaul* or *boka-chaul* on the occasions of some socio-religious functions. In Kamrup or lower Assam, and in the reverine tracts of the valley where *ahu* was largely cultivated, even the higher echelons of the society were also accustomed to that coarse variety of rice. The *Katha Guru Charit* mentions an incident in which Sankaradeva was said to have suffered from diarrhoea due to continuous use of damp *ahu* rice so that his disciple Madhavadeva had to search for *sali* rice from elsewhere in a distant place.[104] It should be mentioned here that the production of *bar* or big, i.e., rough variety of *sali* paddy per acre is usually higher than that of *lahi* or soft or fine variety. Compelled to cultivate in a limited area of land and produce or procure the need for the entire year the *paiks* usually preferred to produce the big variety of *sali* rice to the finer ones.

Storage and volume of production of rice also varied from class to class. While the *paik* peasants had only what they could produce from their allotted piece of lands, the officers, nobles and the kings had huge stores in a number of granaries. For example, Shihabuddin Talish noted that the king possessed more than a hundred and seventy graneries of rice at Gargaon alone besides his stores in other parts of the state.[105] Captain Welsh in 1794 noted that after his occupation of Rangpur (*Sivasagar*) from the hands of the Moamariyas in that year he found large quantities of stored grain, cattle and treasure.[106] The Barphukan with his one thousand *puras* of *rupit* lands cultivated with hundreds of slaves, servants and assigned *paik* workers must have granaries in number. Other dignitaries, the *Satradhikaras* and the priests like Krishnaram Bhattacharyya of the Kamakhya temple who owned almost half of the cultivated lands in Kamrup,[107] must have in the same way, such number of stores. Even officers of the lowest rank like *Bara* who was, in fact, a rich peasant only, is said to have possessed as much as twentysix *puras* of lands, more than hundred workers, about hundred buffaloes, two hundred goats and four cow-shades. While a *Laskar* of the same rank possessed about four to six graneries; a court scholar (*Sabhapandit*) possessed about nine.[108] The rich peasant class consisting of the junior officers and the *Bhuyans* in Kamrup formed a class of 'loose dispersed local gentry', and some of them possessed as many as sixty bullocks, hundred and twenty cows with proportionate number of graneries. It is to be noted that in the petition submitted personally to A.J.M. Mills, the Judge of the Sadar Dewani Adalat, on behalf of the Yuvaraj Ghanakanta, the Charing Raja, Maniram Dewan reported that during the days of the Ahoms, in every village there were two to three such gentry possessing a number of graneries.[109] As against this, if a *paik* with manpower at his disposal possessed 2-3 cows, one pair or two bullocks and a pair of buffaloes, a granery and one or two pieces of silk cloth, he was considered as affluent.[110]

Besides rice and vegetables, innumerable rivers and ponds and dense jungles provided people with fish and animal meat. The *Guru Charit* and aphorisms of Dak bear many references to the consumption of meat and fish.[111] The *Padma Puran* while speaking on the food habits of the people of Assam mentions that they used to take everything except human flesh and they took them without maintaining any ritual.[112] The *Yogini Tantra* also speaks of animals like buffaloes, rhinoceros musk-deer, hair, wild birds, deer, goat, sheep, tortoise, pig, wild fowl and fishes as suitable items of diet even for the goddesses.[113] The Muslim chronicler Shihabuddin Talish also pointed to such food habits of the Assamese people.[114] But all this do not speak of the entire food habit of the people of Assam of our period. The neo-Vaishnavite movement had deep impact on the food habits and other aspects of their socio-economic behaviour.[115] For our purpose it should be mentioned that in Medieval Assam a devoted class of the peasantry being influenced by the ideals of neo-Vaishnavism cultivated abhorrence of meat and showed inclination to vegetarian diets. King Gadadhar Singha is said to have feared that this habit would make them physically weak and mar their military spirit; and therefore, he panalised the Vaishnava preachers who taught this habit.[116] Ralph Fitch also referred to this practice as prevalent in the Koch kingdom, i.e., in lower Assam. The people there, according to him, were "all gentiles (i.e., pure Hindus) and they will kill nothing. They have hospitals for sheep, goats, dogs, cats, birds and for all other living creatures. When they became old and lame, they keep them until they die".[117]

The standard of diet was related to one's class status. While the rich employed professional cook to prepare varieties of items with the use of spices like pepper, clove, cardamom, the poor peasants after the days work could manage to prepare only their bare need.[118] For the Ahom royal kitchen there were the *Changmais*[119] and for that of the Koches the *Supakars*.[120] Other nobles and officers of both the Ahom and the Koch kingdoms, had their own cooks. As a matter of fact while the *paiks* had to be content with ½ *ser* of coarse rice each, the nobles and the aristocrats prepared for them various delicious and expensive items of food. While describing the food preparations made by Chitralekha, the friend of Usha who was married to Lord Krishna's grandson Aniruddha, the poet Ananta Kandali in his book *Kumara Harana* gives a long list of delicious items of the aristocrat classes.[121] The *Darrang Raj Vamsavali* mentions that a Supakar should be proficient in preparing different kinds of dishes.[122] Besides, Sankaradeva in his *Bhagavata-X* gives a detailed list of delicacies for the aristocrats which included among others curd, milk, honey, ghee, rice cakes, fruits, *payasa*, sugar etc.[123] It should be noted that besides use of species like ginger, cumin, *pipali*, oil, chilli, etc. the use of salt (*lon*) was also a mark of aristocratic diet, for salt was very rare in Assam and peasants in general used an alkali preparation called *khar* or *kharani* as a substitute for salt.[124] The term *Khar Khowa Asamiya* used to distinguish as Assamese from other people of the country has been derived from this fact. Salt was so precious that it not only formed an item of gift, but was also compared with gold.[125]

Among other items, fruits of different kinds were used by all sections. But ghee, milk and sugar were rarely used.[126] Ghee was totally disliked by the commonality at Gargaon, the Ahom metropoly,[127] in the middle part of the 17th century when Shihabuddin Talish visited the capital. In a like way, milk was also very rarely used in the peasant society as late as the 19th century. A small amount of sugar was imported from Bengal,[128] but that was used by the richer sections only. Almost all sections of people used to chew areca-nut[129] and betel-vine and it was an indispensable item in all social and religious functions. It appears that those items which were to be bought from outside Assam like salt and sugar including items of dresses and other put-ons could be procured by only the well-off sections of the society and hence the ordinary peasants had no access to use them. They had to produce almost every items of necessity in their own houses; and in that sense, their economy was one of self-sufficiency. As stated by J. Butler "rice, their common food in cheap and abundant; numerous rivers and lakes afford a plentiful supply of fish; their gardens furnish vegetables and fruits, and the climate rendering but little clothing necessary".[130]

Peasants used to put on cloths produced at home. This included a piece of endi or cotton *churia/dhuti* put on above the knee. They had no shirt whatever, but had a *chadar* and a *gamocha* (towel) and had to maintain a strict distinction in their clothings with those of the upper stratum of the society.[131] Women used a *mekhela* on the lower part of the body, and a *riha* in upper woven usually from endi or coarse variety of *muga* or cotton thread.[132] In winter they used a kind of thick and big size cloth made of endi called *Barkapor* or *Jotha kapor*.[133] No body used shoes. There was strict instructions from the state to maintain class distinctions in regard to the quality of dress and nature of wearing. The upper class people used to put on fine varieties of cloths made of *pat, mejankari* and *muga*.[134] Existing remains of cloths put on by the aristocracy show that these were very beautifully woven. Gold ornaments were almost exclusively used by the aristocracy.[135] The ornaments of the peasantry included *keru* (earing made of *jangphai* or ambar) *kharu* (bangles), *anguthi* (rings) etc. usually of silver or copper and rarely of gold.[136] In lower Assam ladies of the families of good birth, i.e., nobles only used to put on *sari*[137] called *Lakshivilash sari* as mentioned in the *Padma Purana* by Narayanadeva.[138] The Ahom aristocracy in later times imported certain items of wear from Mughal India like *pag, jama, chouga, chapkan, paijar* (shoe) etc.[139] which, however, were exclusively used by the aristocrats alone.[140] Thus the peasantry had a distinct identity in the society in their dresses and ornaments. According to Gait "The social distinctions between the aristocracy and the common people and in later times, between the higher and lower castes were rigidly enforced. None but the highest nobles had a right to wear shoes or to carry an umbrella, or to travel in a palanquin, . . . persons of humble birth who wished to wear the *chaddar*, or *shawl* were obliged to fold it over the left shoulder. . . ."[141] But although rigid distinctions were maintained in regard to both quality and quantity of cloths and also in the manner of their wearing, the peasantry in Assam had the needed amount of clothings in both summer and winter. And unlike in some other parts of India where peasants had no cloths than just to hide the secret parts of their body,[142] the peasants in Assam were in an affluent position and had at least three pieces of cloths as their usual put on. The *Katha Guru Charit* points to the dress of an ordinary man as consisting of a *bhuni* (*dhuti*) upto the knees, a *pachara* (*chaddar* or wrapper) and a *tana* or towel.[143] It is to be mentioned that the expertise of the Assamese women in the arts of spinning threads and weaving cloths, reduced the scarcity of cloths among the commoners.

Every peasant family had a small thatched cottage of three or four rooms. There was no house of brick and masonry.[144] Except a few of the kings, all the rulers and the nobles used to reside in the thatched houses.[145] The availability of forest products to supply the entire items of a house might have made the people more adapted to such houses. It is also possible that such houses could resist the havoc of earthquake in the region in which Assam is situated. The two brick masonry houses one at Rangpur called *Talatal* or *Kareng ghar* and the other at Gargaon were built only during the second half of 18th century in the reign of king Rajeswar Singha.[146] There was in fact a prohibitory order preventing even the nobles from building brick houses.[147] It is possibly because of this fact that except for the two brick buildings of our period, remains of other residential brick buildings have not been found anywhere in Assam. Shihabuddin Talish also noticed that in the whole of Assam there is no building of brick, stone or mud, with the exception of the gates of Gargaon and a few temples. Rich and poor alike constructed their houses with wood, bamboo and straw.[148] But there were marked distinctions in style and space.[149] Shihabuddin Talish while visiting the Ahom palace at Gargaon found that the king's audience hall "is 120 cubits long and 90 cubits broad, measured on the inside. It stands on 66 pillars, each of them about four cubits rounds. They have smoothed these huge pillars so well, that at first sight they seemed to have been turned on a lathe. . . . My pen fails to describe in detail the other arts and rare inventions employed in decorating the wood-work of this palace. Probably nowhere else in the whole world can wooden houses be built with such decorations and figure curving as by the people of his country".[150] The houses of the nobles and other rich sections might have followed the same style in accordance with their social status. But the houses of the commoners as described by Butler, "are of the meanest description imaginable . . . a simple hut, ten feet by twenty, divided into a couple of rooms. For sleeping and sitting in, not uncommonly one solitary room, for the only accommodation a man, wife and family possess. . . ."[151] As for furniture and bedstead a small platform of bamboo—two feet high, or in rare occasions a wooden plank called *chalpira* and a *pati* (mat) constituted the bedding. Most of them liked to sleep on the bare ground with a mat as bed.[152]

It therefore appears that the peasant *paiks*, slaves and servants who formed the total labour force of the state and thus the force of productions, had to give away their entire surplus amount of both labour and production to the state and themselves had to live on what amount was left for them as the bare need for maintenance. The statement that the Assamese had very few wants and none of them had even been traders,[153] actually meant their lowest state of survival. In fact, they had to manage their living with what minimum amount of needs they could produce or procure. They were not only economically the poorest section in the society, but were also socially distinct from higher classes. In their food, shelter and cloth—the three basic amenities of life, they were kept in sufficient distance from the well-off section. Material conditions and socio-economic distinctions were also related to castes and family origin. Lower the caste, higher was the burden of work and more rigid was the social distinctions.[154] No man of high birth was to do manual labour of any sort, for his birth was sufficient to make him qualified to be a *chamua*. Members of the subdued Koch, Chutiya and other nobility class were assigned *chamua* status on political ground. But the *paik*-peasants despite their life long labour could obtain this status on very rare occasions only. The result was that the distribution of wealth depended either on caste hierarchy or on political status and very rarely on one's merit or talent.

REFERENCES

1. *Supra*, p. 140.
2. *Report of Captain Welsh* in Mackenzie, *op. cit*, pp. 377ff; Gohain, *op. cit*, p. 115.
3. *Supra*, p. 149f.
4. *Notes by David Scott* in Mackenzie, *op. cit*, p. 380; *HAG*, p. 239; *ADB*, p. 495; Bhuyan, *op. cit*, p. 11.
5. *ADB*, p. 499.

6. *Ibid.*, pp. 437ff.
7. *Ibid.*, pp. 82f; Bhuyan, *Lachit Barphukan*, p. 15.
8. *CP Ins. of Appointment of Katakis at Guwahati, Saka* 1724 (1792 A.D.), *PSN*, Ins. No. 148.
9. The word *paik* was found to have been used in Bengal, Orissa and the Koch kingdom in Kamrup and Goalpara to denote the peasant-cum-foot soldiers. Gait writes that the term *paik* meaning a foot soldier "was formerly very common in Bengal where, for instance, it was applied to the guards who surrounded the palace of the independent Muhammadan Kings" (see *HAG*, p. 239). R.D. Banerjee speaks of the Orissa *paiks* as "descendants of the ancient feudal militia" who "held land in service tenure". According to him, it "was a regular feudal system" " there (see *History of Orissa*, Vol. I, Reprint, Delhi, 1980, p. 25; and Vol. II, Reprint, Delhi, 1980, p. 282). It should be mentioned that the *Darrang Raj Vamsavali* speaks of the *paik* system arranged in decimal order by the first Koch king Biswa Singha (1515-1540 A.D.) (See *DRV*, vv. 196-97). It, therefore, appears that even though the Ahoms had been acquainted with such a system in their habitat in south-east Asian regions (Graham, *op. cit*) prior to their coming to Assam, the complete *paik* structure as they formed here, was more of a local origin. The prevalence of the system of compulsory manual service in the states of Manipur (*lallup*) and Kachar (Brown, *op. cit*, p. 83; Soppit, *op. cit*, p. 27f; *MEAG*, p. 46), lends support to this view.
10. *Asam Bandhu*, p. 74; Rajkumar, *op. cit*, p. 214; *ADB*, p. 498.
11. As a result of the Ahom-Koch friendship established in the first quarter of the 17th century, a number of skilled artisans like *Jugi* (*Katani*), *Tanti* (weaver), goldsmith, *Khanikar* (sculptor) etc. were imported from Koch-Behar to Assam. (See *ABHB*, p. 36; *SAB*, 27; *ABS*, p. 24). In subsequent period a number of Ahom citizens were sent to Koch-Behar for further training in such skills. Assam also imported architects and engineers from outside. One Ghanashyam Khanikar, the architect of the Ahom royal palaces at Rangpur was also from Koch-Behar. (See *TB*, p. 29). Early British writers also noted importation of artisan class to Assam from Bengal as late as the 18th-19th century. See *AAH*, p. 55.
12. J.N. Sarkar, *Freedom Struggles in Medieval Assam*, Dibrugarh University, 1989, p. 19; Bhattacharyya, *op. cit*, pp. 99ff.
13. *ABSM*, pp. 99ff; *ABGB*, p. 84; *Annals of the Delhi Badshahate*, pp. 164ff; *HAG*, p. 157; J.N. Sarkar, *Freedom Struggles in Medieval Assam*, pp. 105ff.
14. *Supra*, p. 142; also fn. 12 above.
15. Gohain Barua, *op. cit*, pp. 3ff; *ABGB*, pp. 175ff. *AAH*, pp. 49ff; Hunter, *op. cit*, I, pp. 32ff; 116ff, 184ff, 240ff.
16. R.S. Sharma, *Perspectives in Social and Economic History of Early India*, New Delhi, 1983, p. 30.
17. *Supra*, pp. 149f. H. Dhekialphukan mentions that junior officers like *Kataki, Kakati, Baragi, Dalai, Ganak, Jyotishi, Khataniyar, Sabhapandit, Bej, Sruti-parha* and *Parsi-parha* etc. were recruited from among the *chamua* class (*op. cit*, pp. 52, 54f). There were, however, hundreds of other officers beginning from the *Baras* to the *Phukans* of such intermediate rank who were also from the same class. See *Mills Report*, Appendix-KB, pp. 610ff; *ADB*, pp. 546ff.
18. *Mills Report*, p. 2; Mackenzie, *op. cit*, p. 7; *DAR*, p. 210.
19. *Mills Report*, p. 2; *Report of Captain Welsh* and *Note by David Scott*, in Mackenzie, *op. cit*, p. 380; *HAG*, p. 239.

20. *Report of Captain Welsh*, in Mackenzie, *op. cit*, p. 380; *HAG*, p. 239, *ADB*, p. 95.
21. *Supra*, p. 51, fn. 58.
22. The first one of a *got* requisitioned for service was called *mul*, the second one *dowal* and the third one was called *tewal*. When requisitioned, the last man was always spared. See *SAB*, pp. 86, 94.
23. *DAR*, p. 201; *HAG*, p. 239; Bhuyan, *op. cit*, p. 11.
24. *Report of the Captain Welsh*, in Mackenzie, *op. cit*, p. 380; *HAG*, p. 239; *ADB*, p. 496; Bhuyan, *op. cit*, pp. 10f.
25. *HAG*, p. 239.
26. Mackenzie, *op. cit*, p. 6; *DAR*, p. 200.
27. The East India Company designated the head of the *khels* as *kheldars*. See Bhuyan, *op. cit*, p. 10.
28. *Report of Captain Welsh* in Mackenzie, *op. cit*, p. 380; *DAR*, pp. 191ff.
29. Wade, *op. cit*, Introduction, p. XXXV; Bhuyan, *op. cit*, p. 10.
30. *Report of Captain Welsh* in Mackenzie, *op. cit*, p. 380; *HAG*, p. 239.
31. *ADB*, pp. 498ff. Gait says that "as members of a *khel* became dispersed in different parts of the country, this system grew most complicated and inconvenient, but it was still in vogue at the time of the British occupation, except in Kamrup where a system of collecting revenue according to local divisions, called *parganas*, had been introduced by the Mahammadans" (*HAG*, p. 240). But the pattern of present villages in Assam in most of which people of the same caste and of the same old professions like weaving, spinning of silk thread, pottery works, black smithy, oil-pressing, fishing, scavenging are found, points to the fact that the system was simplified even prior to the coming of the British to Assam. Also see *Supra*, p. 143.
32. *TB* (Eng.), p. 145; *TB*, pp. 142f; Gohain Barua, *op. cit*, p. 95; *HAG*, p. 239; Bhuyan, *op. cit*, p. 233.
33. *HAG*, p. 239.
34. *Asam Bandhu*, p. 77.
35. *Dakar Bachan*, Ch. XII, v. 129; *KGC*, pp. 55f.
36. *TB* (Eng.), p. 145; *TB*, pp. 119f, 142f; Bhuyan, *op. cit*, p. 233.
37. The rent on the homestead lands began to be imposed since the days of king Kamaleswar Singha. This king also imposed a tax called *barangani* from the *Satras* to meet a part of the accumulated expenditure of the state. See *TB*, p. 138.
38. See *Supra*, pp. 151f.
39. *CHAB*, p. 394.
40. This little amount being the only means of subsistence at Guwahati for those six thousand *paiks*, was pitiably known as *Darrangi Paikar Sambal*. See *Asam Bandhu*, p. 10.
41. It should be mentioned that the system of distribution of lands and labourers among the aristocrats and state-dignitaries in lieu of pay as an aspect of the development of the *paik* system in Assam was inherited by the Ahoms from their original Tai habitats where every state officers possessed lands and labour varying according to his grade. See Hall, *op. cit*, p. 196; George Condominas, 'Essay on the Evolution of the Tai Political System' in *Ethnos*, Vol. 41, Stockholm, 1976, pp. 21ff. The requisition of the local tribes to form the state-labour force and their distribution among the nobles as recorded in the early Assamese chronicles, can be traced back to this original system.

42. See the *Posa* system in pp. 174f.
43. The *Bahatiyas* in the northern hill sides being constantly under the domination of the Daflas, used to imitate Dafla dresses and ways of living in subsequent periods so that there was very little that made them different from their tribal masters. See Bhuyan, *Rajeswar Singha*, p. 238.
44. See for the class hierarchy and their behavioural pattern, *ABHB*, p. 95ff; *ABGB*, pp. 201f; *ADB*, pp. 439f; Gohain Barua, *op. cit*, p. 113.
45. *SAB*, p. 29.
46. *MEAG*, p. 133fn; *CHAB*, p. 410.
47. Also see *Supra*, p. 147.
48. While hiding in disguise and moving in different places of Assam including the *Satras* in his princely life, Gadadhar Singha during the period of crisis in the capital, is said to have noted the wealth of the *Satradhikaras* (heads of the *Satras*) and their kingly ways of living. See *TB*, p. 14. In another occasion the Moamariya Mahanta Nityanandadeva equalised himself in power and status with the Ahom king during Surampha *alias* Bhagaraja (1641-44). See *Buranji Vivek Ratna*, II, (Ms); Bhuyan, *Rajeswar Singha*, p. 140; M. Neog, *Socio-political Events in Assam Leading to the Militancy of the Mayamariya Vaishnavas*, Calcutta, 1982, pp. 31f.
49. J. Bargohain, *op. cit*, p. 26.
50. R.S. Sharma, *op. cit*, p. 62; *MEAG*, p. 107.
51. *MEAG*, p. 111.
52. *Supra*, p. 33fn.
53. *KGC*, p. 560; *HAG*, p. 168; (See Appendix-E).
54. *Mills Report*, p. 3; *DAR*, p. 204.
55. *Asam Bandhu*, p. 77.
56. For the Moamariya rebellion see A. Guha, 'Neo-Vaishnavism to Insurgency: Peasant Uprisings and the Crisis of Feudalism in Late Eighteenth Century Assam' in *The Truth Unites: Essays in Tribute to Samar Sen*, (ed.), A. Mitra, Calcutta, 1985; M. Neog, *Socio-Political Events . . . etc.*; *CHAB*, pp. 301ff.
57. For Moamariya or the Matak kingdom, see S. Dutta, *The Matask and their Kingdom*, Delhi, 1985; R.M. Lahiri, *Annexation of Assam*, Reprint, Calcutta, 1975. For the result see *MEAG*, pp. 122f; *CHAB*, pp. 353ff.
58. *Lakshmi Singhar Buranji* (Ms); Bhuyan, *op. cit*, pp. 270f; *CHAB*, pp. 357, 395.
59. *Lakshmi Singhar Buranji* (Ms); Bhuyan, *op. cit*, pp. 270f.
60. *HAG*, p. 242; *CHAB*, p. 419.
61. *JHR*, IV, No. I, p. 53.
62. *HAG*, pp. 241f.
63. Francis Hamilton, writes that "every man who has a farm, must, in general, work it himself, for labourers . . . can not be procured either for a share of the crop, or for money. The only assistance that can be procured, is that of slaves, and a good many are employed by persons who have influence sufficient to secure property so tangible; and these are chiefly men dedicated to religion, who have lands free of rent". *AAH*, p. 64.
64. S.L. Barua, 'Slavery in Assam', *JHR*, Dibrugarh University, Vol. II, No. I, 1977, p. 75.
65. *DAR*, p. 204. It should be noted that towards the later part of the period under discussion when taxes and other imposts became gradually increased, *paiks* preferred slavery rather than to remain a free *paik*. See *HAG*, p. 241; *JHR*, II, No. I, p. 76.

66. *Document of Buying and Selling of Slaves*, dated *Saka* 1607 (1685 A.D.), *PSN*, Ins. No. 95.
67. *Document of Purchase of Slave, Saka* 1721 (1799 A.D.), *PSN, Bhumika*, p. 72.
68. Cf. *ADB*, p. 501; *ABGB*, p. 210.
69. *KGC*, pp. 101, 470; *Mahapurusha Sankaradeva* by Bhushan Dvij, (ed), D. Barkakati, Jorhat, 1925, v. 637.
70. Cf. S.K. Bhuyan, *Lachit Barphukan and His Times*, Guwahati, 1947, p. 17.
71. M'Cosh, *op. cit*, p. 26.
72. Verse 28, lines 32 respectively. See M.M. Sharma, *op. cit*, p. 279f.
73. See *Supra*, pp. 51f.
74. *PSN*, Ins. No. 150.
75. *DAB*, pp. 13, 15, 52, 55, 109; *AB*, pp. 113, 268; *ABSM*, pp. 8, 27.
76. *JHR*, II, No. I, p. 76.
77. *Guru Charit* by Ramcharan Thakur (ed), H.N. Dutta Barua, 3rd edn, Nalbari, 1978, vv. 3460-3488; *KGC*, pp. 179f.
78. *DAR*, p. 204fn; Gohain Barua, *op. cit*, p. 114.
79. *Report on the Judicial and Revenue Administration, Assam* 1835 (*Letter from Matthie to Jenkins*, August 22, 1835); *JHR*, II, No. I, p. 77, *DAR*, p. 204fn.
80. *JHR*, II, No. I, p. 77.
81. *Asam Bandhu*, p. 77; *ADB*, p. 501.
82. Buying and selling of slave was, however, taken as discreditable in Assam, *Scott to Swinton*, 10 Oct. 1830; Cf. *Report from the Indian Law Commission*, Appendix-VI, p. 317; *MEAG*, p. 53.
83. *AAH*, p. 64; W. Hamilton, *op. cit*, p. 744; *HAG*, p. 242.
84. p. 744.
85. *HAG*, p. 242; M. Neog, *Purani Asamiya Samaj aru Sanskriti*, published in the collection volume *Maheswar Neog Rachanawali*, (ed.), N. Saikia, Dibrugarh, 1986, p. 293. Mathie writes that low castes slaves were sold at a cost about one-third less than the higher caste members of the slave community. See *Report on the Judicial and Revenue Administration*, Assam, 1835.
86. *AAH*, p. 64; W. Hamilton, *op. cit*, p. 744; Khan Choudhury Amanatullah, *Koch Beharer Itihas* (Eng. tr) by S.C. Ghosal, Coch-Behar, 1942, p. 72.
87. *Op. cit*, p. 66.
88. *Report on the Judicial and Revenue Administration*, Assam, 1835.
89. *AAH*, p. 64; W. Hamilton, *op. cit*, p. 371.
90. *Selections from the Records of the Bengal Govt.*, No. XI, Calcutta, 1853, p. 62; *Calcutta Review*, July, 1854, p. 43; *HAG*, p. 242; Gohain Barua, *op. cit*, pp. 114f.
91. Cf. *DAR*, p. 282.
92. *JHR*, II, No. 1, p. 80.
93. *MEAG*, p. 56.
94. *JHR*, II, No. I, p. 78.
95. *Report on the Judicial and Revenue Administration of Assam*, 1835 (*Letter from Jenkins to Mangles*, July 31, 1835); *DAR*, p. 279.
96. *DAR*, p. 279; *JHR*, II, No. I, p. 78.
97. *Report on the Judicial and Revenue Administration of Assam*, 1835; *Letter from Jenkins to Mangles*, July 31, 1835; *JHR*, II, No. I, p. 79.
98. *MEAG*, p. 94.
99. *DAR*, p. 279.
100. *KGC*, p. 101; *Bardowa Guru Charit* (ed.), M. Neog, Guwahati, 1977, p. 113; Nath, *op. cit*, p. 137.

101. *KGC*, pp. 36, 210.
102. *Indian Law Commission's Report*, August 9, 1830, Appendix-VI, p. 317; *JHR*, II, No. I, p. 81. The *Doaniyas* (whose pretended speeches were not understood), who committed much atrocities over the people during the Burmese wars, might have included such robber slaves. See *ADB*, p. 326.
103. *ADB*, pp. 317f; for Burmese plunder see *Weissalisa* (ed), M. Bora, Dibrugarh University, 1977, pp. 27ff.
104. p. 76.
105. *HAG*, p. 134; Sarkar, *loc. cit.*
106. *Bengal Political Consultations*, April 25, 1794, No. 23.
107. *Supra*, p. 53.
108. *KGC*, pp. 51ff, 259, 317.
109. Cf. *Mills Report*, Appendix-KB, p. 606.
110. *Asam Bandhu*, p. 100.
111. *KGC*, pp. 62f, 79f, 100, 105 etc.; *Dakar Bachan*, Ch. VIII, vv. 10-25; Butler, *Travels*, p. 245.
112. *Sristi Khanda*, Chapter 57; cf. N.N. Vasu, *Social History of Kamarupa*, 3 vols. combined, Delhi, Reprint, 1990, p. 71fn; S.K. Chatterji, *Kirata-Jana-Kriti*, 2nd edn, 1974, p. 113. The *Dharma Purana* speaks of the people of Assam as accustomed to beef as well. Cf. B. Kakati (ed.), *Aspects of Early Assamese Literature*, Gauhati University, 2nd edn, 1959, p. 306.
113. Section II, Chapter 9, V. 13.
114. *HAG*, pp. 145f; Sarkar, *loc. cit.*
115. Irfan Habib also has noted that the impact of religion on the food habit of the peasantry of India was great; many of them had to avoid meat from their dietary system. *Habib(B)*, p. 98.
116. *HAG*, p. 169; *ABHB*, p. 60.
117. Ryley, *op. cit*, p. 111f; Foster, *op. cit*, p. 25.
118. In the *KGC* the most minimum preparation of an average peasant is mentioned as *Kharani-marich*, i.e., a kind of preparation with just alkaline and chilli. See p. 254.
119. *DAB*, p. 101; *SAB*, pp. 6f; *ADB*, p. 22.
120. *DRV*, V. 194.
121. *Kumara Harana* by Ananta Kandali (ed.), L. Gogoi, Sibsagar, 1972, p. 41f.
122. *DRV*, vv. 185-86.
123. vv. 11599-11600. These items, however, consisted of the diet of those who belonged to the castes of high ritual status.
124. *HAG*, p. 142; Sarkar, *loc. cit*; *Asiatic Researches*, II, p. 132; *ADB*, pp. 471f; Tavernier gives in detail the process of preparation of alkaline in Assam by burning leaves of the fig trees and vegetable matters called *puni* which grow abundantly in ponds or marshy places. These were first dried and then burnt to ashes. The asheses were then used as alkaline. See *Travels in India*, II, p. 283; P.C. Choudhury, *op. cit*, p. 348.
125. Assamese folk saying.
126. *HAG*, pp. 142, 145f, Sarkar, *loc. cit*; *Asiatic Researches*, II, p. 133.
127. *HAG*, pp. 145f.
128. *AAH*, p. 45. The sugar produced locally (*Supra*, p. 130) might have been of unrefined variety.
129. *HAG*, p. 147; Sarkar, *loc. cit*; *Asiatic Researches*, II, p. 132. Contemporary records show that there were internal trade in areca-nut and betel-leave. *Infra*, p. 237.

130. *Travels*, p. 245.
131. *SAB*, p. 29; *DAB*, p. 102. Butler, *Taravels* . . ., pp. 125f.
132. *ABGB*, p. 201; Dhekialphukan, *op. cit*, p. 107; *CHAB*, p. 420.
133. *AAH*, p. 62; Dhekialphukan, *op. cit*, pp. 106f, 112f.
134. *AAH*, p. 62. The *mejankari* silk, the finest of its kind, procured thrice the amount of the price of the common *muga*. Another kind of silk called *champa* was very fine white silk. Both were put on by the kings and the nobles. See Gohain, *op. cit*, pp. 196f.
135. *SAB*, p. 20; *ABGB*, p. 201.
136. There was restrictions in use of gold ornaments by the *paiks*, slaves and servants. Gold also very rarely became collected at their hands. King Surampha *alias* Bagha Raja issued a prohibitory order preventing commoners to put on gold ornaments. See *SAB*, p. 29.
137. B.K. Barua, *Asamar Loka Sanskriti*, Guwahati, 1961, p. 1500; Neog, *Purnai Asamiya Samaj aru Sanskriti, Rachanawali*, p. 294; Dhekialphukan, *op. cit*, p. 113.
138. Cf. S.N. Sharma, *op. cit*, p. 254fn.
139. *TB*, p. 29.
140. Butler, *Travels* . . ., p. 223; *ABGB*, p. 201.
141. *HAG*, p. 242.
142. *Ain-i-Akbari*, I, p. 389; Ryley, *op. cit*, pp. 118f; *Habib(B)*, p. 102.
143. *KGC*, p. 316; *Guru Charit Katha*, p. 106.
144. *ABGB*, p. 201.
145. Butler, *op. cit*, p. 135; *ABGB*, p. 201; also see *Supra*, p. 109fn.
146. *Asamar Padya Buranji* (ed.), S.K. Bhuyan, 2nd edn, Guwahati, 1964, v. 413; *HAG*, p. 186.
147. *DAB*, p. 126; *SAB*, p. 14.
148. *HAG*, p. 148; Sarkar, *loc. cit.*
149. For the distinctions between the houses of the commoners and those of the aristocrats, and between the kings and his nobles, see *ADB*, p. 493; Gohain, *op. cit*, p. 152; *HAG*, p. 242.
150. *HAG*, p. 150; Sarkar, *loc. cit.*
151. Butler, *op. cit*, p. 135.
152. *Ibid.*; *HAG*, p. 149; Sarkar, *loc. cit.*
153. *Jenkins Report*, in *Mills Report*, p. 59.
154. For instance, a peasant from the Kaivarta community (*Dom*) was killed at the instance of the king Jayadhvaj Singha for his organising a social gathering (*SAB*, p. 31). Gait writes that "Musalmans and Morias, Doms and Haris were forbidden to wear their hair long and members of the two latter communities were further distinguished by having a fish and a broom respectively, tattooed on their forehead (p. 242. Also see *ABGB*, p. 180). The Katanis or the silk thread spinners were despised because their profession was considered as dirty [See *Buranji Vivek Ratna*, II (Ms)]. King Gadadhar Singha selected the Vaishnavas of lower castes like Keot, Koch, Dom and Haris and robbed of their property, forced to eat flesh and swine, cows etc. and compelled them to work as coolies on the roads. See *AB*, p. 271; also Rajkumar, *op. cit*, p. 423.

6

TRADE IN AGRICULTURAL PRODUCTS

Trade in agricultural products grew slowly in Assam in the period under discussion. At the initial stage rulers were busy with maximisation of lands under cultivation in order that the produce could meet the state's demand. It was with this end in view that the entire population was organised and compelled to engage themselves in production works. But there was an automatic limit to the use of their production beyond the level of bare necessity. Their labour and produce both were extracted by the state and a part of these were distributed among the gentry and the nobles so that the surplus left with the peasants hardly encouraged trade in private enterprises. An important blockade in the growth of trade was that under the *khel* system. *Paiks* were bound to do for the state specific works under compulsory corvee system and there was neither liberty of their movement nor occupational mobility. The provision of commutation in money against compulsory manual service, could be hardly materialised. The rate of commutation was also high and the *paiks* and artisans had hardly any cash collections in their hands. Moreover, since the need of armed forces was met from the strength of the peasants alone, the rules maintained a limit in the extent of commutation.[1] As a result the trade of the masses remained almost insignificant. The situation in Kamrup was, however, a different one. Circumstances there permitted an early growth of trade. The Koches, who formed a powerful state in the lower Brahmaputra valley in the early part of the 16th century, developed a money economy. Coins of the Koch king Naranarayan and his successors are found in large numbers and were circulated in the regions of the entire north-east India.[2] To the west, in Bengal and other parts of India, the ruling Mughals who were pursuers of big commercial activities and who had coveted eyes on the natural resources of Assam, had attempted to break Assam's commercial isolation since very inception.[3] Ahom contact with them beginning at least with the 17th century onwards led to gradual monetisation of

Assam's economy in the subsequent centuries. We have mentioned earlier the Koch-Mughal rule over the region of Kamrup prior to its occupation by the Ahoms. This was why, when lower Assam came under the rule of the Ahoms, there had already been trade activities—big and small, and internal and external; and peasants there were well accustomed to the use of money. Thus trade as a form of economy and money as medium of exchange had an earlier history in lower Assam than in upper Assam region. It was therefore, the occupation of Kamrup that became one of the significant factors in the development of trade activity in Assam.

Internal Trade and Items of Exchange

In order to develop a regular trade system, the most needed thing was the volume of surplus production of a state. The voluminous and staple crop in entire Assam in our period was rice. The richer sections who cultivated large plots and engaged numerous workers, gathered enough crop even to make a good surplus. In lower Assam where *sali* crop production was comparatively small and tax burden on the peasantry was also proportionately high, *paik*-peasants could seldom grow or collect sufficient amount of rice even to meet the requirements for a year.[4] But here also a handful of rich peasants and officers collected stores of grains in the form of *guti-bharal* and *gachi-bharal*, which they got cultivated by their slaves, servants and bondsmen. The sufficiency and surplus of rice in upper Assam, however, did not mean growth of any trade in rice in the region. For the peasantry ultimately depended on rice for their living, and even the rich peasants and officers took the constant store of volume of rice as symbol of prosperity and dignity.[6] This was why except in cases of payment of revenue in kind, rice was usually not handled as an item of market. Shihabuddin Talish observed that in upper Assam, peasants had no habit of buying and selling paddy.[7] In other parts of Assam, however, husked-rice—*chaul* formed an item of buying and selling.[8] That rice was accepted as a popular medium of exchange in Kamrup, could be well understood from the fact that even to cross a river by boat, one could offer rice to the boat-man in lieu of money.[9] Many poor people have bartered their other domestic products for rice.[10] Epigraphic records also testify to the sale of rice mentioning its prices. For example, the *Siddheswar Temple Inscription* puts the price of rice at eight *ana* (Rs. ½) per 40 *sers*[11] or one *mon*. But in no foreign or local source rice is mentioned as an article of export. It is true of Fancis Hamilton's account as well which lists all items of export from Assam to Bengal in the early part of the 19th century.[12] It was only in trade with the hill tribes like the Daflas, Miris, Mishimis, Abors, Nagas, Kacharis and Mikirs that rice was exchanged for their hill products.[13] The *Deodhai Asam Buranji* mentions that in a market at Raha in the Mikir and the Kachari border, customs duties included *inter alia* rice,[14] implying that this practice was also applicable to the northern tribes.

Among the cash crops, the most important were mustard seed (*sariyah*) and long and black pepper. Mustard seed was primarily used for extracting edible oil for domestic consumption. Epigraphic records show that oil for burning lamp and as an edible item was bought and sold in the village markets in our period.[15] It was found that oil pressing then became a profession in Kamrup and there was a section of the people who used to extract oil and earned their living by selling it from door to door.[16] The *Katha Guru Charit* speaks of such a professional oil man of Kamrup who in his later life became a Vaishnava.[17] But there is no record to show that oil was produced in such quantity so as to make it an item of export. Mustard seed was, however, collected by the local traders called *bepari* and were exported to Bengal. In the beginning of the 19th century Francis Hamilton reports that Assam had exported annually as much as 15,000 *mons* of mustard seed to Bengal.[18] Within the kingdom also there were some small local markets called *Sariyah-hat*, where mustard seeds were largely sold.[19] Peppers were used as spices and as medicine. Its production was considerably high and its entire surplus in upper Assam was collected by the *mudoi* traders of Kamrup,[20] who exported it to Bengal. Hamilton records that it was through the Assam choky at Goalpara that an amount of 50 *mons* of long pepper and 50 *mons* of black pepper were exported to Bengal per annum.[21]

Among other domestic products sold in the markets mention may be made of areca-nut and betel-leaves, pulses (*mah*) like *mug and mati* and jaggery etc. In the *buranjis*, however, there is no reference to these items as sold in the markets. But that areca-nut and betel-leaves formed universal items of trade throughout Assam can be known from both literary and epigraphic records. Shihabuddin Talish who is almost silent about trade in upper Assam, speaks of buying and selling of areca-nut and betel-leaves even in the heart-land of the Ahom capital at Gargaon. 'The only traders who sit in the bazar', he says, 'are betel-leaf sellers'[22] The *Guru Charits* also speak of markets in the region round about the capital. It is said that a lady, the mother of a Vaishnava saint called

Bhabanipuria Ata, managed a daily shop at Nazira-hat and used to sell goods contained in three to four pairs of baskets (*bhar*) every day.[23] From epigraphic records, it is learnt that one *guchi* (20 leaves) of *pan* (vine) usually cost 21 cowries, and one *pon* (80 pieces) of *tamol* (betel-nut) 1 *ana*.[24] There were, however, such poor people who had not even that amount of cowries to buy one *guchi* of betel-leaves.[25] In any case, trade in areca-nut and vine which could bring only small amount of profit was confined to small traders and *paik*-peasantry as it did not involve large scale capital investment.

In Kamrup *matimah* cost 5 *gandas* (4 cowries = 1 *ganda*) per *ser*, *mug* 40 *gandas* per *ser*, and *gur* 1 *ana* for one *ser*.[26] To the contrary, in Gargaon as Shihabuddin Talish records in 1662, *mugdal* and vetch (*mati*) were used to be sold for Rs. 10 and Re. 1 each per *ser*.[27] But that was an unusual time and local people when compelled to sell them, might have demanded such soaring prices. It should be mentioned that while vegetable items like brinjal are recorded to have been sold in Kamrup in certain circumstances,[28] record of sale of vegetables in upper Assam are not available to us. Other items of trade included cloths of various qualities and sizes. In Kamrup cotton towels (*gamocha*) were available in the village markets and the price of a *gamocha* was 4 *pons* of cowries.[29] Production of cotton cloths grew as a profession all over the state during our period. In Kamrup particularly, certain places like Soalkuchi, Hajo, Lakharpara of Rangia, and North Guwahati, in upper Assam Narayanpur were famous for fine and skillfully embroidered fabrics.[30] The *Tantis* or the weaver class peasants of Kamrup were well-known for their products. Some of them used to weave cloth on some other's behalf by taking wage at the rate of one *pon* of cowri per cubit.[31] The Koch king Naranarayan is said to have organised the weavers-guild at Tantikuchi in Barpeta where varieties of cotton cloths were produced.[32] A major part of the cloths were sold out. Besides cotton, silk cloth—*pat*, *muga* and *eri* (endi) were also produced and a part of these were sold in the markets. Hamilton records the prices of these cloths as follows:[33]

Pat silk	Rs. 1 to 8 (as per size and quality)
Mejankari silk:	
Dhuti	Rs. 1 to 4
Mekhela	Rs. 1 to 3
Chadar	Rs. 2½ to 4
Mosquito curtain	Rs. 1 to 6
Riha	Rs. ½ to 3

Eri:

Riha	4 to 6 *anas*
Cheleng	8 *anas* to Rs. 2
Cotton:	
Dhuti	Rs. 10 per piece

(finest one measuring 12 cubits by 3 cubits)

According to Francis Hamilton, small dealers used to collect the items from peasant families and "hawk them about from house to house, for there were no shop and few markets (for cloth)".[34]

The state had to export some items of value to Bengal to meet the need for salt and certain items of dress and luxury for the upper stratum of the society. Shihabuddin Talish noticed that formerly "once a year, by order of the Raja, a party used to go for trade to their frontier near Gauhati, they gave gold, musk, aloe wood, pepper, spikenard and silk cloth in exchange for salt, sulphur and certain other products of India. . . ."[35] The export-import relations of Assam particularly with Bengal are evidenced by all contemporary records. In the beginning of the 19th century, it is reported that silk thread and silk-cloths were exported to Bengal.[36]

Towards the later part of our period imported salt was sold in internal markets in small quantities. The inscriptions recording the items and expenditure incurred in connection with the maintenance of two temples make mention of the prices of salt as 1 *ganda* per *tola*.[37] In the internal markets, poor peasants were sometimes found selling edible ferns called *dhekia* and fuel wood (*khari*).[38]

Various kinds of garden fruits including banana, oranges, mangoes, pineapples etc. were common articles of sale in the markets.[39] Shihabuddin Talish mentions that at Gargaon during Mughal occupation, oranges used to be sold at ten for a *paisa*.[40] Occasionally some other items like *chira* (flat rice), sugar, etc. were also sold in small quantities.[41] Buchanon records that during 1808, 1,000 *mons* of ghee, and sugar worth Rs. 1,000 were imported from Bengal.[42] Ghee was mainly used by the priests for worship and lightening lamp in the temples and by the people coming from outside Assam as an edible item. In Assam, it was neither produced in large scale nor was it consumed by the commonality most of whom had such an abhorrence of ghee that they would not eat any dish which had a flavour of it.[43] As a consequence, when the Mughal soldiers at Gargaon had to buy ghee, it was Rs. 14 per *ser*.[44] Even milk was hardly sold in upper Assam and as late as the 19th century there was no *goala* (milk-man) caste to buy, sell and produce milk and milk-items on a professional basis.[45]

Besides agricultural and its ancillary produce like cloths of silk and cotton, items of bamboo-work like *jakai, khaloi, pachi, kharahi, dala, pala, kula* etc. were sometimes sold.[46] Horses, bullocks, cows, goats, domestic and wild birds like ducks, pigeons, *tuni* bird and fishes sometimes formed items of trade.[47] It was found that horse was very costly in the north-eastern region and sometimes old at Rs. 70.[48] Artisan classes like potter, blacksmith and goldsmith might have sold a certain amount of their produce to obtain other necessaries of living. There is a reference in the *Satsari Asam Buranji* to the buying of earthen pot (*malacharu*) by the *paiks* spending two cowries for each.[49]

The price of the commodities, however, did not remain static. It varied sometimes region wise and sometimes as per the contemporary situation, and sometimes as per quality. Price of commodities was higher in Kamrup than it was in upper Assam. For example, while a *gamocha* was sold in Kamrup for 4 *pons* (4 *anas*) of cowries, the same cost 1½ *ana* in upper Assam.[50] Price rise was also due to famine, drought and internal instability and as Shihabuddin Talish records even *matimah* at Gargaon was sold for Re. 1 per *ser* in 1662, an item which cost only 20 cowries in Kamrup in peace times.[51] During the Moamariya rebellion production and population ran short and prices of even essential commodities rose to an unprecedented level. For instance, in the period after this rebellion when a great famine occurred, one lime pot of rice (about 250 grams) costs Re. 1 and one mango Re. 1 to Re. 1-8 *anas*.[52] In normal times, price was, however, low all over Assam. The comparatively higher rate of prices in Kamrup was due possibly to the repeated Mughal invasions, population density and commercialisation of products.[53]

Hills-Plains Exchange and Medium of Transaction

Trade with the neighbouring hill tribes was basically an exchange of hills and plains products. For example, the Mishimis brought a kind of poison known in Assam as *Mishimitita,*[54] an aromatic herb called *gathiyan*, special kind of sword and spears, while from the Singphos Assam received ivory, copper and silver.[55] In exchange they took from the market place at Sadiya, rice, glass-beads, cloths, salt and money.[56] The Abars and the hill Miris exchanged pepper, ginger, manjit (madder), wax and *jim* cloth (cotton blanket) for rice, fish and other plains products.[57] Similarly the Nagas, the Garos, the Mikirs, the Khasis took from the plain's market, rice, duck, cattle, goat, cloth, salt, dried-fish and tobacco for cotton, chilli, ginger, yams and iron implements.[58] Butler records in 1837 that the Mikirs and other hill tribes brought down their cotton to sell or barter for salt.[59] Thus in the market place where the hills people collected plains products, the medium of exchange was naturally barter.

In the markets and shopping corners of the plains, trade was carried on both through money and barter system. Contemporary records, literary and epigraphic, speak of the existence of coins of different denominations.[60] Shihabuddin Talish mentions that the currency in Assam consisted of cowries, rupees (silver) and gold coins.[61] Extant hoard of Ahom coins, however, donot contain gold coins in large number. Even silver coins of early Ahom rulers are markedly absent.[62] These are found in sufficient number only since the time of king Siva Singha (1714-1744). The existence of cowries as stated by Shihabuddin Talish is, however, corroborated by all contemporary literary sources. The epigraphic records stating prices of the items in terms of *kara* or cowries is a positive evidence in this case. The *buranjis* and the Vaishnava biographical works contain references to the use of cowries in daily transactions.[63]

In the lower Assam region Koch coins called *Narayani* or *Narani* or *Narengi* were in circulation since the middle of the 16th century.[64] These were of one rupee and half rupee denominations.[65] The *Charit Puthis* and the Koch genealogical works make mention of the use of money in transactions. For example, the *Darrang Raj Vamsavali* records that the Koch king Naranaryana granted twenty five thousand rupees for the maintenance of the Kamakhya temple.[66] The biographical works of the Vaishanava saints speak of *taka* or *rupiya* (one rupee silver coin), *adtaka* (half rupee coin), *siki* or *maha* (quarter rupee), *admaha* or *duani* (1/8th of a rupee) and *kara* or cowries.[67] Some of the contemporary Assamese chronicles refer to huge capital in terms of money invested by big merchants.[68] Besides, references to the use of cowries in small transactions are found in contemporary epigraphic records also wherein oil, rice, betel-nut and vine are said to have bought and sold in terms of cowries.[69]

Frequent mention of money in transactions, however, did not mean its wide circulation and distribution among all classes of people. The *paik*-peasants were not at all well off to make stock of silver coins. The instances of one peasant who was said to have been over serious about his only savings of rupees five, of another for rupees fourteen,[70] and still two others who had mortgaged themselves to a rich peasant for rupees five[71] show that cowries and a very few silver coins were the usual deposits of the people at large.[72] As late as the 16th century such scarcity of coins could be marked. It is recorded that once Madhavadeva, the apostle of Sankaradeva, could not get the change for five rupees in smaller denominations even by searching for the whole day.[73] That silver coins were rarely used in small transactions by the peasants, is proved by what was recorded by Ralph Fitch who stated that the people in lower Assam (in Koch kingdom) used almond, a kind of fruit, as a medium of exchange.[74] The use of cowries in small and everyday transactions like those of areca-nut, betel-vine, fuel-wood, vegetables, mustard oil, towels, fishing appliances etc. indicate extensive use of cowries among the common people. Against this, there were very few well-off persons in the society belonging usually to the higher sections, who had large collections of silver coins. The *Katha Guru Charit* speaks of a man who had savings of four hundred silver coins.[75] The merchants who traded in almost in all parts of the north-eastern region could invest largest amount of capital.[76] Badan Chandra, the Barphukan of Guwahati, could offer as many as sixty thousand rupees to get that assignment for himself[77] and gave lots of pearls, ornaments, garments, furniture, slaves and servants as dowry to his daughter.[78]

Market System

The rulers of Assam since early times established some market places in important junctions of communications between the hills and the plains people for exchange of their products. The Persian chronicle *Tabaqat-i-Nasiri* by Minhas-ud-din Siraj mentions as many as thirty five passes in the north providing communication link between the hills and the plains. One of such passages was the famous market place of Karampatan where, according to the author, was a cattle market held everyday, and where about one thousand and five hundred *tangahan* horses of Bhutan were sold.[79] There was also an important market at a place called Geegunshur on the Assam Bhutan border where the Bhutanese exchanged silver bullion to the amount of two lakh rupees, a considerable quantity of gold dust, cowtails and woollen cloths for rice, silk, iron, lac, deer skin and buffalo-horns etc.[80] The Ahom rulers encouraged the growth of such hills-plains trade centres on the border areas. An important centre of such trade activities was Sadiya in the extreme north-east corner of the kingdom where almost the entire tribes of the north-eastern hills came to exchange their goods.[81] Other centres facilitating the hill tribes to trade with the plains were those of Nagora and Kacharihat in Sibsagar where the Nagas from the hills and the Kacharis from the plains exchanged their products.[82] In Kamrup, Kukuriya within the Barduar area was an important trade centre visited mainly by the Garos.[83] Suhungmung and his son Suklenmung founded a number of markets in places like Bakata, Sala and some other places on the banks of the river Dihing and Dikhow.[84] King Pratap Singha and later king Rudra Singha and Rajeswar Singha established markets at places like Titabar, Namchang, Dopgarh, Barhat, Raha, Gobha, Phulaguri and Sonapur.[85] Towards the close of the Ahom rule, two markets were established at the town of Dichai, now known as Jorhat, on the banks of the river Dichai (Bhogdai) by *Rajmantri* Purnananda Buragohain.[86] Besides, in every important town, there was a market place (*hat*) like those at Guwahati, Nazira and Hajo. Stephen Cacella, the Portuguese traveller, who visited Koch Hajo about the year 1627, refers to Hajo as one of the most important and populated towns of the time.[87]

In the meantime shop keeping largely developed in Mughal India.[88] Stephen Cacella mentions that in Koch-Behar also there were many *bazars* and markets.[89] It is, therefore, natural that shop keeping as means of trade developed also in Hajo which was for a long period under the Mughal administration. References are, however, lacking about the existence of shops in other parts of Assam. The *Katha Guru Charit* speaks of rural traders as packing up their goods in bamboo baskets and hawking them in the rural markets (*pahar*).[90] It was only after the expedition of Captain Welsh in 1792-94 that some accompanying tradesmen opened a few shops in Guwahati.[91] This had inspired some enterprising local people to take up this business in the years that followed. During Chandrakanta Singha's reign (1811-1822) three shops were established at Jorhat by local traders.[92] Marketing in Assam by women was not unknown in the period under review, as we have earlier mentioned, but it were the Muslim women only who used to keep small shops in their own residences.[93]

Besides *hats* or markets, fairs or open air markets were held in some occasions of important socio-religious festivals in different places of the kingdom. These included the *Ambubachi mela* at Kamakhya (3rd week of June), *Ashokastami* fair at Soalkuchi and Silghat (Nagaon), *Maghipurnima* fair at Udalguri in Darrang, fair at *Phukanhat* in Lakhimpur, and the Buddha-*purnima* fair at Hajo. In all the fairs marketing continued for several days together.[94]

Means of Carriage and Trade Routes

The most convenient and usual way of carriage was by boats. While referring to trade activity, contemporary literary sources speak of big boats laden with goods and belonging to the traders. The *Katha Guru Charit* contains numerous references to river-borne trade. In one place, it speaks of such a heavily loaded boat that it had every possibility of getting sunk in the river.[95] It also talks of a big merchant who traded in Assam, Bengal, Garo hills and Bhutan regions chiefly by boats.[96] The Brahmaputra with all its tributaries provided the river way between upper and lower Assam, and between Assam and its neighbouring states, chiefly Bengal. It is learnt from the *Tripura Buranji* that there was river borne traffic between Assam and the Kachari kingdom. A trader from Sibsagar could go *via* the Namdang, the Brahmaputra and the Kapili upto Demera within the Kachari kingdom. The entire journey took a period of twelve days.[97] The Mughal traders carried their trade with Assam entirely in boats through the Brahmaputra. One Ratan Shah Mudoi is said to have secretly loaded his boats in the Brahmaputra with Assam products.[98] M'Cosh noticed that many of the boats were "of enormous size and capable of carrying one or two hundred maunds (*mons*), though cut out of the solid trunk of one single tree. . . . When bulky cargoes, such as cotton, are brought down the country, a common practice is to fasten two canoes together with transverse beams, so that the canoes remain three or four feet apart; the platform is then loaded with cotton or straw, and in this way they admit of carrying a much larger bulk than they could when ununited and with much greater safety".[99]

It was a fact that river traffic in trade was not free from danger; though only on rare occasions boat sank causing loss of life and property. M'Cosh says, the "voyage up the Brahmaputra is attended with many obstacles. . . . In the dry season there are no beaten paths to facilitate tracking; the boatmen must either force their way through the high reeds on the crumbling perpendicular bank or scramble along the bottom; or what they prefer, keep upon the shoal side of the river, where the said bank affords good footing, though with the great drawback of the boat's getting often a ground".[100] Strong current in the river in the summer, wind, storm and even sand slowed down the movement of the boats. Stephen Cacella writes that big trading boats from Gaur, Rajmahal and Patna moved up and down the Brahmaputra throughout the whole year and it was a very tedious journey taking 25 to 35 days from Goalpara to Calcutta and about 33-45 days upstream.[101] In the same way journey through the Brahmaputra within Assam proper was also time consuming. However, to avoid danger of river traffic, the traders moved in groups and formed temporary partnership to do their business in distant lands.[102] Earlier the course of the Brahmaputra near Hajo had a sharp curve and it was full of sands which caused much difficulty to the movement of boats. The Koch king Naranarayan is said to have straightened the course of the river by getting its bottom excavated in about 1564 A.D.[103]

The boats were built considering several factors like weight, size and durability. There were different types of boats like *holong* (hollowed out from a single piece of log), *petela-nao*[104] (boats which were very wide in the middle), *khel-nao* (boats used for sports) and *charanao* (boats used by the aristocracy and the nobility) etc.[105] The last two types of boats were beautifully designed, richly decorated and painted with lime, vermilion, yellow arsenic and indigo.[106] Such boats were used for war purposes also. The *Katha Guru Charit* speaks of different types of big boats like *konhra, manchai, kondh, pinish, cherengadari, bojra* etc.[107] Persian sources like the *Baharistan-i-Ghaybi* and the *Fathiya-i-Ibriya* speak of another type of boat called *bachari* used by the Assamese for war purposes.[108] The latter work mentions that most of the boats in Assam were built of *chambal* wood, "and such vessels, however heavily they may be loaded, on being swamped do not sink in the water".[109] According to it in the year 1662, in the month of *Ramzan* about 32,000 such boats owned by the Assamese frequented Guwahati.[110] There was a professional caste called *Maji*, who specialised in boat sailing to distant lands some of whom were also traders. In a like way the *mudois* who traded in boats had also carried passengers to distant lands on payment of money (known as *kharcha* or *chukani*).[111]

Alongside, trade through overland routes was carried on. Through the mountain passes communication with the neighbouring countries like Bhutan, Tibet, China, Burma etc. was maintained. With Bengal there were, besides water-ways, three land-routes through Goalpara.[112] There were also routes connecting Guwahati with Cachar, Sylhet and the Jayantiya kingdom.[113] In case of land-routes, however, there were problems of carriage. The manpower force which this system needed had also formed a part of the peasantry and the militia. As a matter of fact, at the time of war and harvest parties to carry goods of large amount were hard to find. The small traders, including the female counter-parts, however, carried on their heads or shoulders the baskets of their goods.[114] Heavy or large quantity of goods which could not be carried on heads, backs or shoulders, as a single load, were divided into two and tied into two ends of a bamboo bar called *dang*, or a flat bamboo piece called *biriya* or *baoka*. The *dang* along with the loads was called *bhar*. When the goods were too heavy to be carried on by a single man in *bhar*, it was done by two men by binding the loads in the middle of a strong bamboo-pole, and depending on the weight of the load, even by four men by putting the goods on a square or rectangular size flat bamboo appliance bound to two strong bamboo poles called *sangi* or changi.[115] Like the *banjaras* in Mughal India who organised the transport of goods and their regular movements through pack-oxen[116] there was no such professional trader-cum-carrier class of men in Assam on land routes in our period. Haliram Dhekialphukan in the early part of the 19th century observed that in Assam there was no bullock cart; goods were carried here by individual men on their backs and by boat.[117] This was because, only few Assamese traders could go beyond the border for trade. In lower Assam, however, on special occasions bullocks were used to carry goods.[118]

Transportation of goods to distant lands was not only conditioned by lack of good overland routes or of man-power carrier, but also by geographical and topographical conditions of Assam. Consequently distant trade was the business of rich people only who could collect labour and take the risk of riverine trade. Small traders, therefore, confined their business to their own locality only.

Contemporary records speak of different bamboo appliances as containers of commercial goods. These included baskets of different sizes like *duli, pachi, tom, japā* etc.[119] Earthern pots like *kalah* and *tekeli* were also in use.[120] Goods like cotton were bound in form of bales, and sugarcane, betelvine etc. were in form of bundles.

Classes of Traders

Traders were variously called depending upon the nature and volume of their trading commodities. Thus the *sadāgar* or *sāud* and the *mudoi*—both were big traders who carried on trade through river between upper and lower Assam, and also in the neighbouring states. Small traders were called *bepāris*, and when female they were called *pohāris*.[121] There were *mudois* who dealt in some specific items like gold, silk-cloth and silk, salt, mustard-seeds, elephant-tasks etc. and sometimes even in areca-nut. The *Katha Guru Charit* speaks of merchants who traded in salt and betel-nut.[122] This work classifies big merchants into three categories: (i) those dealing in gold and jewels, (ii) those dealing in different types of silk, and (iii) those who traded in articles like salt and alkaline called *khār*.[123] It, therefore, appears that specialisation or monopoly in trade also grew to certain extent, leading to the emergence of a professional class of people called *Māji* who took the responsibility of safe carriage of goods by boats.

Towards the later part of our period money lending system on interest also grew. According to the *Katha Guru Charit* on one occasion, a trader of Kamrup region who borrowed Rs. 120 was to repay Rs. 500 inclusive of both the capital and its interest after 7 years,[124] the rate of simple interest being 59.1 per cent in that case. Sometimes small traders having failed to make profit abandoned the venture for good and tried their best to avoid contact with the lenders who frequently demanded repayment of the loan with interest.[125] The situation was however different in upper Assam. Here goods were abundant and cheap[126] and trade as mentioned above did not develop as such. This was why the traders of Kamrup looked on upper Assam as "something like—paradise for merchants", where all valuable trade commodities could be procured in good quantities and at considerably low price.[127]

Trade Regulations

Development of trade and market system brought towards the later part of our period, a considerable amount of revenue to the state treasury. It was possibly with this idea of increasing revenue that the Ahom kings specially since the days of Pratap Singha were taking measures to regulate and promote trade. It should be mentioned that the border markets were sometimes raided by the neighbouring hill tribes. The rulers of Assam, therefore, established police outposts and customs houses called *choky* or *duar* in order to check the raids and regulate imports and exports; and to collect the customs duties civil and military officers, usually of the ranks of the *Baruas*, were put in the charge of these outposts. The biggest of such *chokies* or *duars* was the Asam-choky variously known as Hadira-choky, Kandahar-choky or Banghalhat which was placed under the *Duaria Barua*.[128] It was his duty to receive goods from Assam merchants and to pass them to Bengal and *vice versa*. He also levied the customs duties on all imports and exports and collected an annual income of about Rs. 90,000 in average although out of this amount a sum of Rs. 26,000 only finally reached the state treasury.[129] The *Duaria Barua* at Kandahar-choky was empowered also to receive in advance goods from Bengal which were considered as of immediate need.[130] In the same way the *Duarias* in charge of the other *chokies* also acted as customs officers and collected the state share of the goods exchanged. Rules framed for regulating and collecting state's share were first introduced by king Pratap Singha. It is recorded in the chronicles that the collection for the state treasury from any market place was ¼ of each kind of goods brought for sale.[131] Calculated at this rate, Mills in his report of 1853 gives the following abstract of income of the Ahom kings collected from some of the important trade centres or market places of Assam during 1813-14.[132]

Market	*Income per Annum (Rs.)*
Bangalhat	60,000
Batakuchi	6,000
Solaphat	6,000
Gobha, Sonapur and Raha *hāts*	6,400
Guwahati and other *hāts*	12,000
Namchang	5,000
Jorhat	11,000

The rate of taxes on markets, was not static; it varied from time to time according to price and demand of the commodities. In an earlier chapter we have mentioned, for example, the various rates of levies collected at the market at Raha in the Ahom Kachari border.[133] Besides, from the income of the market places on the border areas, a part was given to the respective hill chiefs or kings. It is recorded in an Assamese chronicle that from the collection of the market places on the Kachari border a part was earmarked for the king of the Kacharis.[134]

The growth of market places and levy of varieties of taxes on the goods sold, and on the places and routes used towards the later part of our period, indicate a gradual growth of an organised system of trade in the period under review. The entire trade with outside states like Bhutan and Bengal was carried on by some big merchants who used to collect goods through smaller traders from various parts of the kingdom and finally exported them to outside the state. It appears that the traders from lower Assam who had expertise in the art, used to collect goods in upper Assam at cheaper rate and sold them in other places with much profit. There were also merchants who tried to evade taxes; but they were panalised when detected.[135] On the other hand, there were provisions for allowing tax-free trade, called *ādāni* which were usually permitted to those who traded for the income of the temple or other religious institutions.[136]

Trade in Assam was not so developed to be compared well with that of Mughal India. There was no organised money lending community or system to issue of *hundi* or bil of exchange and any *bimā* or insurance system in Assam which were working so well and extensively in Mughal India.[137] As stated above there grew only in a very limited way, lending and borrowing of money system in individual level.[138] Repayment of debt along with interest was, however, made obligatory both by conscience associated with superstitious belief and the state administration.[139] Even now almost all the villagers of Assam believe that one who does not repay his debt, shall in the next birth, be an ass to bear the burden of the lender. If the borrower had no means to repay the debt, he was compelled to do physical labour as a bondsman in the lender's house. According to R.S. Sharma, the exploitation of physical labour of the debtor was considered to be an effective legal way of collecting debts, in ancient India also.[140] Monopoly, in the true sense of the term and in a bigger way, as in Mughal India,[141] did not grow in Assam. Officers appointed by the state controlled the check-posts and registered exports and imports and only the registered merchants were allowed to trade in other lands on condition of paying taxes.[142] An Assamese chronicle mentions that one such merchant called Jayhari traded in the Jayantiya Kingdom investing a capital of Rs. 8,000 with the permission of the king and the prime minister.[143] Such big merchants sometimes specialised in certain goods like salt, cloth, mustard-seed etc. as we have mentioned earlier.

The backward state of trade in Assam in our period can be guessed from the fact that emphasis on production of cash crops, (like tobacco and indigo) was not given here as in Mughal India.[144] As for the price of the commodities, it was cheapest in upper Assam, cheaper in lower Assam and cheap in Bengal in comparison to all other parts of Mughal India.[145] Prices also varied according to the quality of the goods. For example, while the price of a *tola* of gold of good quality was 18-20 rupees that of an inferior quality Rs. 12-14 only.[146] Again while the price of a *muga* cloth of a certain size was 1-4 rupees that of a *mejankari* of equal size was Rs. 2.50 to 6.[147]

The growth of trade had an important impact on the life style of the people. It was through trade with Bengal that people got acquainted with some of the finest products of India, especially sugar, varieties of silk and fine cotton dresses, perfumes etc. and also salt which were not available in Assam. Growth of trade provided the poor peasants to earn their living by selling some small items they produced. Their earning was, however, very little and in most cases major part of it was expended for paying revenue particularly in Kamrup. So, money hardly got collected with them. At the same time, a small but richer class grew in the society. They were big merchants who made much profit by collecting the local goods, in cheaper rate and selling them outside the state in higher prices. It was thus the possibility of having sufficient scope of a profitable trade in Assam that attracted the attention of the British in the 18th century, who first concluded a commercial treaty with the king of Assam in 1793 A.D.[148]

REFERENCES

1. *MEAG*, p. 111; Also see Dhekialphukan, *op. cit*, p. 52.
2. H.E. Stappleton, 'Contributions to the History and Ethnology of North Eastern India' in the *Journal and Proceedings of the Asiatic Society of Bengal*, New Series, Vol. VI, 1910, p. 158; *PSN, Bhumika*, p. 7; D. Nath, 'Currency in the Koch kingdom: Problem of the Koch Coins', in the *Journal of Indian Numisnatics and Sigillography*, Vol. IV, Indore, 1988, pp. 157f.
3. Evidence show that there were secret trades between Assam and the Mughal India carried on by the Mughal traders since the 17th century. The chronicles record interesting references to such trades in which even officers of ranks as high as the *Barbarua* of Assam were associated in order to get articles of luxury produced in Mughal India. Ahlier Khan who was the Mughal Faujdar at Guwahati during 1638 is said to have presented king Pratap Singha with a special necklace of pearls in order to get elephants from Assam. The offer was, however, rejected by the king. See *PAB*, pp. 87f.
4. The *KGC* which reflects in the most, the social life of the people of the lower Brahmaputra valley, particularly of Kamrup, frequently speaks of scarcity of food—*akal*, when the peasants had no grain of food to maintain their family. See pp. 469, 478, 564 etc.
5. *Ibid.*, p. 437.
6. There is a folk saying in Assam that those who have no rice/paddy, deserve no respect in the society. (*Yār nāi dhān, tār nāi mān*).
7. *HAG*, p. 150; Sarkar, *loc. cit.*
8. The biography *KGC* (pp. 469, 478) mentions that in times of scarcity, even money could not buy rice, implying that in usual times rice was bought and sold.
9. *Ibid.*, p. 504.
10. *Ibid.*, p. 331. Also see *MSMCD*, v. 1250-51.
11. *PSN*, Ins. No. 24 (Addend), p. 180; (See Appendix-B).
12. See Appendix-A.
13. *Supra*, pp. 174f.
14. p. 142.
15. See, *Supra*, p. 130fn.
16. *KGC*, pp. 243, 461. Although oil was extracted from mustard seed almost in every family, professional *Telis* used to supply oil to the royal house-hold and

the aristocracy. Such *Telis* were permanently attached to those establishments. See *ABHB*, p. 124.

17. pp. 243, 461.
18. *AAH*, p. 46; see Appendix-A.
19. *Buranji Vivek Ratna*, II, (Ms). As recorded in the *KGC* Gopal Ata, one of the Vaishnava preachers of medieval Assam, was a trader in mustard seed. See p. 87.
20. Upper Assam was recorded in the Vaishnava biographical works as a land of *agar* wood, black pepper, gold, silk cloth (mulberry), elephant tasks, decorated wicker, i.e., *japi* etc. The work *KGC* mentions that *mudoi* traders from Kamrup came by boat through the Brahmaputra and collected mustard seed from the houses of the peasants in Upper Assam. See p. 87.
21. *AAH*, p. 46; Appendix-A.
22. *HAG*, p. 150; Sarkar, *loc. cit.*
23. *KGC*, p. 250.
24. *PSN*, Ins. No. 24; (Addend), pp. 178f; (See Appendix-B).
25. *KGC*, p. 82.
26. *PSN*, Ins. No. 24; (Addend), pp. 178f, pp. 17f. For sale of *gur* (jaggery) see *KGC*, p. 366; Gohain, *op. cit*, p. 166.
27. Cf. J.N. Sarkar, *The Life of Mirjumla*, p. 322.
28. *KGC*, p. 237.
29. *KGC*, p. 87.
30. *MSMCD*, vv. 270-71; *MEAG*, p. 110; Gohain, *op. cit*, p. 185; Nath, *op. cit*, p. 150f.
31. *KGC*, p. 393.
32. *Ibid.*, p. 208; *Guru Charit* by Ramcharan Thakur, vv. 3606-3607.
33. *AAH*, p. 62.
34. *Ibid.*
35. *HAG*, p. 144; Sarkar, *loc. cit.*
36. Among the import from Bengal items of luxury included fine pulses, stonebeads, corals, woolens, Banarasi silk cloths, tafetas, gold and silver embroidered cloths, muslin etc. See *AAH*, pp. 45f.
37. *PSN*, Ins. No. 24 (Addend.), pp. 178ff; (See Appendix-B).
38. *KGC*, pp. 123, 307. The *buranjis* record that the Maran and the Barahi tribes supplied the Ahom royalty with these articles free of cost. *SAB*, pp. 6f, 10; *DAB*, pp. 101f, 105.
39. *KGC*, pp. 542ff; *HAG*, p. 142, Sarkar, *loc. cit.*
40. Sarkar, *loc. cit.*; B.K. Barua, *op. cit*, p. 95; *Habib(B)*, p. 101fn.
41. *PSN*, pp. 178ff; *KGC*, p. 366.
42. *AAH*, p. 45.
43. *HAG*, p. 146; Sakrar, *loc. cit.*
44. Sarkar, *op. cit.*; Cf. Sarkar, *The Life of Mirjumla*, p. 322.
45. *AAH*, p. 63.
46. *KGC*, p. 331.
47. *Ibid.*, pp. 261, 289; Gohain, *op. cit*, p. 166.
48. *Kachari Buranji*, p. 46.
49. p. 94.
50. Cf. S.L. Barua 'Market System in Assam during the Days of the Ahoms', *Assam Economic Journal*, Vol. III, No. I, Dibrugarh University, 1977, p. 66; *Supra*, p. 238.

51. *Supra*, p. 238; *PSN*, Ins. No. 24 (Addend), p. 180; (See Appendix-B).
52. *TB*, p. 111.
53. Also see *Assam Economic Journal*, Vol. III, p. 66.
54. The Snake Pillar Inscription of Sadiyo, Cf. K.N. Dikshit, *Annual Report for 1924-25*, Archaeological Survey of India, pp. 157f.
55. *DAR*, pp. 243f; Butler, *op. cit*, p. 122.
56. *Ibid.*, p. 243; *Ibid.*
57. *Ibid.*
58. *Ibid.*, pp. 242, 244; *AAH*, p. 45f; *Jenkins Reports in Mills Report*, pp. 207ff.
59. *Travels*, p. 22.
60. *KGC*, pp. 54f, 60, 87, 91f, 94, 149, 289, 341; *MSMCD*, vv. 1250-1251; *PSN*, Ins. No. 24, 75. Also see the two *CP Inscriptions of the Dirgheswari Temple, Saka* 1661 (1749 A.D.) and *Saka* 1679 (1757 A.D.), pp. 184ff.
61. *HAG*, p. 143, Sarkar, *loc. cit.*
62. *HAG*, p. 242. The earliest of the Ahom coins now extant, belong to the reign of king Jayadhvaj Singha (1648-1663). See *The Comprehensive History of Assam*, II, p. 19.
63. *SAB*, p. 94; *DAB*, pp. 142, 148; *ABS*, p. 27; *KGC*, pp. 82, 87, 92 etc. Also see E.A. Gait, 'Notes on some coins of the Koch Kings', *JASB*, 1895, Vol. LXIV, Pt. I, No. 3; Stappleton, *loc. cit*, pp. 155ff.
64. Nath, *loc. cit*, p. 156; *PSN*, *Bhumika*, p. 7.
65. See Stappleton, *loc. cit*, pp. 155ff.
66. v. 551.
67. *MSMCD*, vv. 1250-1251; *KGC*, pp. 62, 149, 160, 289, 341, 497. King Gaurinath Singha (1780-1795) issued coins as small as 1/32th part of a rupee. The present author possesses such a coin of Gaurinath Singha.
68. *Jayantiya Buranji*, p. 19; Wade, *op. cit*, pp. 46f.
69. See *PSN*, Ins. Nos. 17 and 24, (Addend), pp. 177ff; also *DAB*, p. 142; *KGC*, pp. 103, 110.
70. *KGC*, pp. 494f, 497.
71. *Ibid.*, p. 128.
72. Nath, *loc. cit*, p. 158.
73. *KGC*, p. 91.
74. Ryley, *op. cit*, p. 113.
75. p. 343.
76. *KGC*, p. 85.
77. The Barphukan is said to have collected this amount through various means including raiding of gold and silver articles of the Kamakhya and the Hajo temples. See *TB*, p. 131; also Bhuyan, *op. cit*, p. 390; S.L. Barua, *op. cit*, p. 224.
78. *Asam Buranji* (Ms) by Padmeswar Singh Naobaicha Phukan; S.L. Barua, *op. cit*, p. 227.
79. Raverty, *op. cit*, pp. 567f.
80. Martin, *op. cit*, p. 662; Pemberton, *op. cit*, p. 83.
81. *Report of Captain Welsh* in Mackenzie, *op. cit*, p. 387; *DAR*, p 243; Gohain, *op. cit*, pp. 167f; *Assam Economic Journal*, III, No. 1, pp. 63f.
82. Pemberton, *op. cit*, p. 182; *Assam Economic Journal*, III, No. I, pp. 62f.
83. *Ibid.*
84. *ABS*, pp. 17f; *Buranji Vivek Ratna*, II (Ms); *SAB*, p. 21.
85. *DAB*, p. 142; *SAB*, pp. 177, 183; *Buranji Vivek Ratna*, II (Ms); *ABHB*, p. 40; *Jayantiya Buranji*, pp. 17f, 127.

86. *Buranji Vivek Ratna*, II (Ms).
87. Wessels, *op. cit*, p. 123.
88. J.N. Sarkar, *op. cit*, pp. 121ff; Chopra, *op. cit*, pp. 34f.
89. Wessels, *op. cit*, p. 128.
90. *KGC*, pp. 249f, 366, 545.
91. *ABGB*, p. 189.
92. *Ibid.*, p. 189; *CHAB*, p. 433.
93. *ABGB*, p. 189.
94. *Mills Report*, pp. 157ff. Appendix-M; Wade, *Geographical Sketch*, p. 13; *Assam Economic Journal*, III, No. 1, p. 63; *SHODHAK*, Vol. 15, Pt. C.No. 45, p. 114.
95. p. 250.
96. *KGC*, p. 85, 89, 94.
97. *Tripura Buranji*, (ed), S.K. Bhuyan, 3rd edn, Guwahati, 1983, p. 21.
98. *DAB*, p. 110; *PAB*, p. 63; *SAB*, p. 80.
99. M'Cosh, *op. cit*, p. 28.
100. *Ibid.*, p. 7.
101. Wessels, *op. cit*, pp. 128f; Bhuyan, *op. cit*, p. 55.
102. See *KGC*, p. 85; A. Guha, 'Medieval North-East India: Polity, Society and Economy 1200-1750 A.D.', Occasional Paper 19 of the Centre for Social Science Research, Calcutta, 1978, p. 17.
103. *DRV*, vv. 450-51.
104. *PAB*, p. 63; *DAB*, p. 110.
105. Dhekialphukan, *op. cit*, p. 108f; Gohain, *op. cit*, p. 202.
106. Such a boat (a *khelnao*) has been preserved in the Assam State Museum at Guwahati.
107. pp. 185, 187, 353, 360.
108. *Baharistan*, I, pp. 243, 247; *HAG*, pp. 147f; Sarkar, *loc. cit.*
109. *HAG*, p. 148; Sarkar, *loc. cit.*
110. *Ibid.*, p. 147; *ibid.*
111 *Guru Charit Katha*, pp. 53, 55, 229; *KGC*, pp. 59f, 91.
112. M'Cosh, *op. cit*, p. 9; *DAR*, pp. 245ff.
113. *DAR*, pp. 245ff.
114. *KGC*, p. 250.
115. *Chungrung Phukanar Buranji*, pp. 6f; *Jayantiya Buranji*, pp. 99, 101; *KGC*, pp. 250, 470, 542; *PSN*, Ins. No. 137.
116. *Habib(B)*, pp. 66f; Sarkar, *op. cit*, p. 129; *The Cambridge Economic History*, I, p. 331.
117. *Op. cit*, p. 104.
118. Ram Saraswati who was the court-poet of the Koch king Naranarayan, and who translated the *Mahabharata* into vernacular (Assamese) mentions pairs of bullocks (*baladhijora*) carrying books to the poet's library. See *Pushpa Harana Vanaparba* (of the *Mahabharata*), vv. 3936-37. About use of bullocks to carry books see also *KGC*, p. 207.
119. *Jayantiya Buranji*, pp. 99ff; *Changrung Phukanar Buranji*, p. 6ff.
120. *Ibid.*
121. *SAB*, p. 90; *KGC*, pp. 366, 545.
122. *Mudois* dealing with salt were called *Lon-mudoi*, and those who traded in betel-nut were known as *Guwa-mudoi*. See *KGC*, pp. 65, 94, 553.
123. *Ibid.*, p. 492f.

124. *Ibid.*, p. 553.
125. *Ibid.*
126. *Ibid.*, p. 89.
127. Neog, *Sankaradeva and His Times*, p. 79. Also see *KGC*, Introduction, p. V.
128. The kings of Assam appointed the *Duaria Barua* sometimes by issuing properly written royal orders. One of such orders written on a copper plate has been collected by the Assam State Museum in 1970 from the Deputy Commissioner's office of Kamrup. See *PSN*, Ins. No. 154, pp. *Bhumika*, 103f. Sometimes two persons were appointed to hold the same post. See Bhuyan, *op. cit*, p. 50; *Supra*, p. 181.
129. *Welsh to Cornowalish*, Feb. 28, 1793; *Bengal Political Consultation*, March 11, 1793, No. 17; Bhuyan, *op. cit*, p. 341.
130. *ABGB*, p. 189.
131. *ABS*, p. 38; *SAB*, p. 29. Maniram̃ Dewan states that traders in valuable goods like *muga* cloth were to pay as *dan* ½ of the total amount of sale proceeds. See Maniram Dewan's Report to Major Adam White, in B. Sharma, *Maniram Dewan*, Guwahati 1950, Preface, p. iv, viii.
132. See Appendix-K.B., p. 621.
133. *Supra*, p. 179.
134. *DAB*, p. 142.
135. *KGC*, pp. 85f.
136. *KGC*, p. 85; *The CP Inscription of Hayagriva-Madhava Temple Endowing Boats etc., Saka* 1705 (1783 A.D.); *PSN*, Ins. No. 144.
137. *Habib(B)*, pp. 74f.
138. *KGC*, p. 553.
139. This might have been believed in line with the theories of the ancient Indian social thinkers like Narada. See *Dharmakosha*, (ed), Lakshman, Shastry Joshi, 1937-41; Cf. R.S. Sharma, *op. cit*, p. 215.
140. *Op. cit*, p. 215.
141. *Habib(B)*, p. 85f.
142. Once a trader called Jayhari was detained and his goods confiscated in Jayantiya kingdom owing to his wrong selection of route. It was only through the good office of the king of Assam that he was finally released. See Wade, *op. cit*, pp. 46f.
143. *Ibid.*
144. *Habib(B)*, p. 87.
145. *Ibid.*
146. Wade, *Geographical Sketch*, p. 20.
147. *AAH*, p. 62.
148. Aitchison, *op. cit*, pp. 112f.

7

CONCLUSION

Despite it's being handicapped in many other ways to build the high road of progress, the medieval period witnessed significant changes in many aspects of the agrarian life of Assam. Before the coming of the Ahoms and their settlement in the extreme south-eastern part of the Brahmaputra valley in the early 13th century A.D., the region was almost entirely inhabited by some aboriginal tribes like the Marans and the Barahis. Their main source of living was shifting cultivation of rice and collection of wild fruits, roots, herbs, and catching of fish and hunting. The incoming Ahoms with a comparatively advanced knowledge of agriculture and economy found it easy to subjugate these tribes and utilise them in the production process of the state that they had built up here in which most of the local shifting tribes became settled agriculturists.

The act of agrarian expansion in Assam in the post-13th century period was also accelerated by the experience the ruling Ahoms gathered from other neighbouring tribes and castes, who had already developed some sort of economy and formed states in the Brahmaputra valley. Some of them, e.g., the Kacharis and the Bhuyans and those who had already raised themselves to higher caste-echelons were familiar with plough and wet rice culture of higher out put. But their agriculture-based economy being predominantly characterised by shifting variety of cultivations, they failed to successfully defend themselves against the incoming Ahoms, who had brought with them a superior technology and the idea of a well planned and developed agrarian system.[1] Under the circumstances the Ahom rulers found it easy to follow a policy of gradual land acquisition in the territories they conquered and to convert these into large cultivable fields by systematic engagement of local labour force to the pursuit of agriculture. Thus the Ahoms opened a new chapter in the agricultural history of Assam in which both the land and its people became the objects of primary interest of the state.

Although ideologically the system of considering both land and men—the latter irrespective of birth, as properties of the state, remained the same till at least the last part of the period under study, it was remodelled from time to time to suit certain internal and external situations. Internally, the most significant factors which gave a new direction to this policy were the acceptance of Brahmanical Hinduism by the Ahom rulers and the consequent inclusion of caste hierarchy as a *faith accompli* in the agrarian social structure. In due course large scale gift of lands were made to the Brahmanas and religious institutions along with allotment of *paiks* or labourer class of people for their services. This had resulted not only in the creation of a landed class in the society on the basis of caste, but it also led to the birth of a master-slave social structure. Allotment of specific professions to specific sections of the people classified on the basis of birth, exactions of compulsory manual service from them, and exemption of certain groups of people from this obligatory service at the same time on the basis of caste, was to a large extent a product of this social innovation. Gradually caste factor came to be associated with the process of production relations and also with class structure or social stratification.

The important external factor came out of the need for the defence of the state specially against the Turko-Afghans and the Mughals from the west, who invaded Assam during the period times without number. This needed a strong military base and sufficient economic strength for which changes had to be made in the agricultural system. After their contact with the Mughals, the Ahom kings adopted many elements of administration of the Mughals, particularly in the field of land survey and land measurement and switched over to money economy in the subsequent period. This was how, the Mughal pressure broke Assam's economic isolation of self-sufficiency and opened the way for the growth of trade and commerce with other parts of India. To this result may be added another the coming of the British to Assam and their participation in its politics. The Moamariya rebellion and the failure of the Assam government to quell it with the traditional *paik* militia, and at the same time, the success of the British standing army to fight the rebels, convinced the ruling class to abolish the feudal militia obtained through the *paik* system and consequently to switch over to keeping a standing army. As a result, Assam's agrarian economy during the medieval period was in a stage of transition first from tribalism to feudalism[2] and then from feudalism to the advent of modernism.

Once land and people became established as the properties of the state, the question of their distribution and ownership also came along with. The present study shows that the Ahom state very carefully took census of the people and surveyed the lands, and distributed them between different classes of people in different proportions. While the king, the members of the aristocracy and the nobility, and those associated with administration, took the lion's share in terms of both man and land, the priestly class, and those associated with religious functions, were also given a considerable part of these wealths. The interesting aspect of this distribution system was that since the population were regarded as the property of the state, they (i.e., the peasants) were safeguarded from being exterminated by any factor through the grant of certain amount of land sufficient to produce the least amount to maintain a family. Consequently, there emerged inequality in distribution of lands among different sections of the people. While lands in bigger plots concentrated in the hands of the state dignitaries, the aristocracy and the ecclesiastical class, the peasants got the minimal share. Thus the privileged class enjoyed double advantages: they could reclaim lands in bigger plots, and retain them as their hereditary property. At the same time paik-peasants could neither possess big farms like those of the *khats* or gifted lands, nor retain as hereditary property whatever lands they obtain from the state or themselves had opened by clearing jungles. But then, there was the guarantee that those who actually tilled the land, must have a plot of it.[3] Indeed, in the Ahom state there was hardly any peasant or working man except the slaves who had no lands. For this guarantee of having a cultivable piece of land, a peasant was to render his compulsory service to the state for three to four months in a year in whatever from the state demanded of him as per his proficiency even though that might be against his will. Thus depending upon the necessity of the state every individual capable worker automatically became a peasant. But the point to note is that he had only the right to cultivate the land but not the right to handle it as personal property unlike the owners of the *khats* and the gifted lands.

Distribution of land and exploitation of the peasants were somewhat different in the lower Brahmaputra valley.[4] The Ahoms found in upper Assam almost a virgin land and a community of people to be formed or reformed according to their need; in lower Brahmaputra valley which in early times formed the centre of the kingdom of Pragjyotisha-Kamarupa, the case was quite different. The society of this region was greatly influenced by the Brahmanical system, and as such it had experienced feudal structure of administration at least since the middle of the 5th-6th century A.D. The system of payment of salaries to the officers in terms of lands as also men and gifting of lands to the Brahmanas and religious institutions along with labourers,[5] gave it a feudal structure even in those early centuries. During the 14th to 16th centuries almost major part of the area was under the supreme control of the Bhuyans who also largely helped in feudalising the society.[6]

Subsequently the Koches, with certain characters of the Turko-Afghan and Mughal administration of Bengal[7] had ruled this part for about a century when the Mughals occupied it and formed it a province of their empire. It was during this period that the Mughal system of collection of cash revenue from the peasants through the intermediaries like *Choudhury* and others[8] was adopted and they were authorised to collect from the peasants their land revenue on the king's behalf. It was under the circumstances that when a peasant failed to pay his part of the revenue, he was usually forced to pay it.[9] But there in Kamrup, as in Mughal India, peasants possessed proprietary rights over the soil, and therefore, could sell or mortgage it. Again, as a result of the existence of a number of inter-mediaries associated with revenue collection, a large part of the peasants' production was alienated from them so that no more was finally left with the peasants beyond the requirements of their subsistence.[10] When the Ahom rulers occupied Kamrup, they did not abolish this established structure. In stead, keeping the old system almost in tact, it was provisioned that land and properties of the revenue officers in Kamrup were to be increased proportionately to the amount of revenue they could actually deposit to the state treasury.[11] And naturally the revenue officers in Kamrup became so rich that even a part of the total land possessed by a *Choudhury* was something around one hundred *puras* of cultivable land.[12] This might have adversely affected the peasantry some of whom either left lands and accepted slavery as means of living, or mortgaged their labour to richer classes for money to pay revenue. This was why contemporary records of our period speak of landless peasants, land-sale, wage labourer,[13] bondsmen and even existence of greater number of slaves in Kamarup. As a result, when in upper Assam every peasant possessed a piece of land and had collection of paddy to meet the needs of his family for the whole year, there in lower Assam, were peasants who might be landless or had accepted bondage as a way of living. This was why while poverty was not unknown in Kamrup, peasants in upper Assam had possessed at least the bare needs of the family except in times of famine and drought.[14] This fact was at the root of an ununiform agrarian life throughout the Upper and Lower Assam region.

In both upper and lower Assam, however, there was imbalanced distribution of cultivable lands and their ultimate ownership remained with the state although in Kamrup peasants enjoyed hereditary proprietary rights over the soil.[15] But then, the peasants were the men who fought for the defence of the land, who worked for the non-manuals and it were they who provided leisure and pleasure for them. Under such circumstances, as Andre Beteile says, "the agrarian society of south Asia (and as such of Assam) are generally marked by great inequalities of wealth, power and status. . . . The high and low born, the rich and the poor, those for whom others work and those who work for others differ from each other in almost every significant way. They live separately, dress and speak differently and divide their time differently between work and leisure".[16]

The Ahoms brought important changes to Assam's agrarian life in the period under review in many other ways. The introduction of *sali* rice and embankment system revolutionised Assam's agriculture. It was this *sali* rice that gave the actual strength to the state to gradually expand at the cost of the neighbouring tribes and thereby to integrate a greater part of the valley into one political entity and to defend it against the mighty Muslim invaders. The period is also significant in the field of land extension and its improvement. Clearing of dense jungles grown in favourable topographic conditions and bringing them gradually under cultivation by way of survey, resurvey, settlement and distribution among the peasantry and improving their quality by construction of embankments, made Assam a flat level plain full with growing crops which caused amazement even to the foreigners coming to Assam.[17] To this might be added the topographic as well as the Mughal factor. While the former made the lands normally fertile, the latter contributed to its proper and useful managements. It is to be mentioned that development of agriculture through proper management and utilisation of lands by way of survey and measurement was to a great extent, a legacy of the Mughals.

Despite this, the agrarian economy of Assam in our period did not follow the Mughal pattern characterised by its extensive trade and industry, production of large amounts of cash crops and dependence of the state on money economy. This could be explained in the light of Assam's geographical location which cut it off from the main land of India on one hand, and the existence of hills almost on all sides on the other, which prevented its easy access to Mughal India. The *paik* system with its all rigidity in peasants' mobility led to marked absence of professional guilds,[18] and kept Assam in a backward economic position. Above all, till at least the last decade of the 18th century the Ahom rulers preferred to follow a policy of isolation almost like that of the contemporary Chinese rulers[19] due to which Assam could hardly conceive an idea of the developments that had already taken place in other parts of contemporary India.

It appears, however, that the economic structure of Assam would have been changed to the Mughal pattern had the Mughals continued their rule for at least another century or had expanded it to the eastern Brahmaputra valley. The fall of the Mughals in the beginning of the 18th century negated the need of keeping an ever-ready military force and a strong economy in Assam for its defence. At the same time internal strife and the civil war in the last part of the 18th century leading to the British participation in Assam politics, and external aggressions led by the Burmese ultimately brought the final collapse of the Ahom kingdom.

The period under discussion, however, failed to bring about a radical change in the agrarian social structure in all its aspects. The role of caste and the outlook towards the menial class as reflected in the Hindu social laws, ware rather encouraged than discarded. In the society the menial castes were always regarded as somewhat inferior. The lot of the Kaivartas, the Katanis, the *Talis*, the Dhubis, the Haris and other such castes did not improve.[20] Distribution of land and labour, and ownership of lands in most cases depended upon caste or birth. This was how, agrarian life, its social outlook as it was in other parts of India,[21] or as it germinated in the kingdom of Kamarupa in early times, remained the same during our period also. The so called low caste people remained with the same status, pursuing the same professions till at least the close of our period. It is to be noted that the civil gentry who formed an intermediary class and depended for their existence largely on assigned lands, domestic labourers or allotted workers in lieu of salary, were usually recruited from among the members of the higher echelons of the caste and status hierarchy. Thus the *Phukans, Baruas, Bujarbaruas, Duaria Baruas, Rajkhowas, Choudhurys, Kakatis* etc. who were economically sound, were either from the higher caste or from the higher status. Even the Muslim factor in which social equality is largely recognised,[22] and which certainly added some new elements to the agrarian life of Assam of medieval period, failed to break this social structure and its production relations and the Neo-Vaishnavite Movement of Sankaradeva and others.

Thus the agrarian life in Assam in medieval period contained certain demoralising elements like—concentration of lands in the hands of certain sections through reclamation and gifts, acquisition of their proprietary ownership over the lands thus acquired, exercise of overall power by the holder to handle the lands and peasants therein, creation of a compulsory militia from among the peasants, existence of a servile landless class to serve the well-off sections, payments of salary to the state officers in terms of land and labour, collection of revenue as much as possible in both cash and kind, primarily with the objective of consumption of the privileged class, a secluded and self-sufficient economy with little scope for trade among the peasantry worth the name, and above all, lack of the peasants' individual ownership of cultivable lands. All such aspects gave some feudal characters to the society where inequality in both social and economic aspects was so prominent that it could be called a class-society.

The feudal system (as may be put as semi or quasi-feudal) with its strong demarcation between different groups, gave rise to peasants' protest in course of time. The Neo-Vaishnavite movement led by Sri Sankaradeva and a host of his disciples in the 16th and the 17th centuries propagating equality of all sections of people, at least in the spiritual plane, also had a share in the growth of such protests. As a result, in our period of discussion we have seen that peasants sometimes protested against heavy taxation as in the cases of the peasants of Darrang and Nagaon; sometimes they rose in armed rebellion under the banner of religion getting united through ethnic ties against the hatred and exploitation, as was the case with the Moamariya rebellion. It should be noted here that the Moamariya rebellion in which most of the downtrodden peasants were associated, had far reaching consequences in subsequent history of Assam. Although for a short time, the assumption of royalty by a commoner by dethroning the Ahom monarch during the rebellion broke the age-old faith of the people on the divinity of the throne.[23] The commoner, now onwards, began questioning the existing politico-social inequality in the society derived on the basis of noble birth or political status. For example, the use of the *dola* or palanquin which was a symbol of aristocracy in our period, now began to be demanded by the peasants also and the same was conceded to them by the state as well. They also obtained the right even to disapprove the appointment of the great Gohains[24] which was otherwise not their business at all. Towards the close of our period peasants even opposed the system of compulsory military service.[25] Thus, the psychological unpreparedness of the peasantry to serve the state under compulsion in armed forces with age-old weaponry and organisation, proved finally ineffective for the defence of the state and the ruling class itself became convinced to abolish this out-dated system[26] and switch over to a new one which has already symbolised the dawn of modern age in other parts of the world. But the entire history of peasant protest leading to movements in Assam does not fall within the purview of our discussion.

The period was, however, one of transition and experiment in the field of agrarian economy. Improvement of agriculture through the exploitation of land and man power resources, establishment of a self-sufficient economy and its gradual break through the entrance of money and trade, attempt to recognise social equality of the people in the spiritual level[27] at least were some of the salient features of the period. Above all, it finally marked the abolition of exaction of compulsory manual service from the peasantry through the *paik* system. The system being highly unpopular from the time of its very inception, failed to work, when the revolt of the Moamariyas exposed the inner weakness of the Ahom monarchy. As a result, the peasant-*paiks* were found to have openly refused to serve the state as soldiers during the last part of the Ahom rule. This was why there was no alternative for the state to continue its existence than to abolish the system itself. It was, in this sense, that this period was one of transition and experiment.

REFERENCES

1. From the nature of the sites selected by Sukapha in the Brahmaputra valley prior to his establishment of the capital at Charaideo in present Sibsagar district, and the use of buffalo pulled ploughs in wet rice cultivation, J.N. Phukan has rightly come to the conclusion that the Ahoms were a valley dwelling people rather than hill men as pointed out by Gait. See Phukan, *op. cit*; and *HAG*, p. 70.
2. *MEAG*, Chapter 5.
3. There was no chance in Assam under the Ahom state system for the growth of a landless proletariat as pointed out by J.N. Phukan (see *op. cit*). But that was true only more particularly in case of upper Assam, and not in the lower Assam region.
4. According to J.N. Phukan, it was a compromise between the Mughal and the Ahom systems. *Ibid.*
5. For early land grants in Assam see M.M. Sharma, *op. cit*; *JHR*, Vol. IV; R.S. Sharma, *op. cit*, pp. 1f, 62.
6. *MEAG*, pp. 82ff; *India Past and Present*, III, No. 2, pp. 243ff; *CHAB*, pp. 193ff.
7. Nath, *op. cit*, p. 128.
8. *Supra*, pp. 154ff.
9. *Habib*, *op. cit*, pp. 190ff.

10. *Ibid.*, p. 190.
11. *Supra*, p. 155fn.
12. Cf. Bhuyan, *Rajeswar Singha*, p. 260.
13. The *Guru Charit Katha* (p. 87) mentions a class of wage labourer who used to reap in other's field in times of harvest for a part of rice they reap. They were called *duniā*, for they used to take one *don* of rice for reaping for a day.
14. Scarcity of food occurred in Assam in the period under discussion in many times for causes like flood, drought, ravages of war etc. But cases of famine resulting in large-scale depopulation, as in Mughal India (See J.N. Sarkar, *op. cit*, pp. 270f) except in times of external aggressions and civil wars, are found to have occurred on only few occasions. The chronicles mention that large scale extermination of people due to famine occurred in Assam in 1642-43 and 1789 A.D. See *AB*, p. 126; *DAB*, p. 84; *TB*, p. 111.
15. Major references to land sale in our period are found only in case of lower Assam. Records of land sale in upper Assam are few and far between. See *Supra*, pp. 66f; J.N. Phukan, *op. cit.*
16. Andre Beteille, *Studies in Agrarian Social Structure*, 3rd Impression, New Delhi, 1979, p. 38.
17. *HAG*, p. 141; Sarkar, *loc. cit.*
18. *CHAB*, p. 395.
19. K.S. Latourette, *The Chinese—Their History and Culture*, 4th revised edn., London etc., 1972, pp. 450ff. H.M. Vinacke, *A History of the Far East in Modern Times*, First Indian edn., New Delhi, 1978, pp. 294f.
20. See the role of Gadadhar Singha in selecting the peasants of lower castes and to compell them to work and partake meats of the uneatables. *AB*, p. 271. Also see *Supra*, pp. 18ff.
21. Irfan Habib, 'Presidential Address' in the *Proceedings of the Indian History Congress*, Kurukshetra University, 1982, p. 26.
22. *Alberuni's India* (tr.) Edward C. Sachau, London, 1910, Vol. I, p. 100; Habib, 'Presidential Address', p. 25; M. Neog, *Sankaradeva And His Times*, Gauhati University, 1965, pp. 360ff.
23. *HAG*, p. 149; Sarkar, *loc. cit.*
24. *Supra*, p. 197.
25. *TB*, p. 112.
26. Realising that feudal militia would no longer serve the interests of the state, Prime Minister Purnananda Buragohain in the year 1794 abolished the peasant militia and introduced a standing army in the state. See *AB*, p. 356; *SAB*, pp. 171f.
27. The leaders of the Neo-Vaishnavite movement propagated equality of all sections of the people—the *Brahmanas*, the *Sudras*, the tribes, women, workers, artisans and the like, on the ground that the soul of everybody including that of the animals is none but the God Himself. As a result it was in the religious gathering that peoples of different caste and class could assemble inside the same house known as 'namghar'. But then, in practical life there are grounds to believe that distinctions between man and man of different categories were maintained. For example, *Katha Guru Charit* mentions that while an ordinary *bhakat* was not permitted to put on a dignified dress, the same was allowed to the members of the spiritual lords. The fact that, in neo-Vaishnavite ideal the leaders put the *Guru* above all and regarded him as possessed of all supreme powers, points to the development of the conception of a feudal-lord in the society indicating the

existence of an all powerful class over a powerless one. This, again, points to a master and slave society. See, H. Gohain, *Kirtan Puthir Rashbichar*, Gauhati University, 1981, p. 26; and *Asamiya Jatiya Jivanat Mahapurusia Parampara* (Guwahati, 1987) for details.

APPENDICES

APPENDIX-A

Export-Import List Between Assam and Bengal 1808-09

Imports to Assam from Bengal

Salt at 5½ rupees	1,92,500
Fine Pulse	800
Ghiu (1000 mans)	1,600
Sugar	1,000
Stone beads	2,000
Coral	1,000
Jewels and Pearls	5,000
Cutlery and glass-ware (European)	500
Spices	1,000
Paints	500
Copper	4,800
Red Lead	1,000
English Woolens	2,000
Tafetas	2,000
Benares silk cloth, called Kinkhap	500
Satin	1,000
Gold and silver cloth	1,000
Shells	100
Muslin	10,000
	Rs. 2,28,300

Exports from Assam to Bengal

10,000 mans of stick lac	35,000
Muga silk, 65 mans	11,350
Muga cloth, 75 mans	17,500
Manjista or Indian madder	500

Black pepper, 50 mans	500
Long pepper, 50 mans	300
Cotton with seed, 7,000 mans	35,000
Ivory	6,500
Bell-metal vessels	1,500
Mustard-seed, 15,000 mans	20,000
Iron hoes	600
Slaves, 100	2,000
Thaikhol fruit, 50 mans	150
	Rs. 1,39,900

Source: AAH, pp. 45, 46.

APPENDIX-B

A Part of the Text of the
Land Grant Copper-plate Inscription of the Siddheswara Temple
Saka, 1645 where prices of different commodities are recorded.

(See for Assamese Language in Original Copper plate).

Source: Deputy Commissioner's Office, Kamrup, Transcript No. 247, *PSN*, pp. 177-80.

APPENDIX-C

A part of the Text of the
Copper Plate Inscription of Devottara Land Grant towards Daily Puja of the Umananda Temple, *Saka* 1661 which records grant of different classes of servitors.

(See for Assamese Language in Original Copper plate).

Source: PSN (Addend), pp. 181-82.

APPENDIX-D

Relevant Part of the Text of the
Copper Plate Land-Grant Inscription to Ramananda Vidyalankar, ***Saka*** 1669, which records highest amount of land donated through a single inscription.

(See for Assamese Language in Original Copper plate).

Source: Itihase Sowara Chasata Bachar, S. Rajkumar, p. 393.

APPENDIX-E

Relevant Part of the Text of the

Copper Plate Land and Servitors Grant Inscription to Brahmachari Satra, Nagaon, *Saka*, 1722 which among others records the system of land measurement.

(See for Assamese Language in Original Copper plate).

***Source: PSN*, Ins. No. 97, pp. 82-83.**

ENGLISH TRANSLATION

Earlier to this it was represented by Viswambhara Brahmachari of Nagaon to king Lakshmi Singha that when Vasudeva was removed to a place called Cheuni, 60 of his *bhakats* (bachelor devotees) were forced to take the duties of *kanris* or *paiks*. The king placed this matter for disposal with the Chief of the Judiciary, Sivaram Barbarua of the Adabariya Sandikoi family, and (Kekeru) Kalita Phukan. These officers after proper scrutiny of the whole business ordered the cancellation of registration in regard to 10 *gots* of *paiks* (out of the total of 20 *gots*) in Nagaon and reinstated them as *bhakats* with grant of some land under the service of Viswambhara. Viswambhara now preferred an appeal to the Prime Minister, Purnananda Buragohain, to record the gift on a *tamrapatra* (copper-plate). The Buragohain forwarded the matter to the king and by his orders, sent Dhanbar Bara to Nagaon to make an enquiry into the matter. The Bara held his trial among the people of Telia and Khatowal *khels* in Nagaon. These people gave their evidence in writing that Lakshmi Singha had given such men and lands to the Brahmachari. On the Bara's report the Prime Minister issued the present copper-plate writ, bestowing on Viswambhara Brahmachari 9½ *gots* of *paiks* and 124 *puras* of lands. To the east and west of the land is Telia, to the north Panigaya, and to the south Prasadkhowa. The length of the plot measures 620 *beon* and its breadth 80, amounting to 124 *puras* of land. Within this campus is a homestead plot 80 by 80 *beons* amounting to 16 *puras*. Viswambhara Brahmachari and his descendants are entitled through this copper-plate to enjoy this land for generations by performing *sravana-kirtana* congregation and by blessing the Svargadeo, i.e., the king. Nobody will be allowed to do otherwise. *Saka* 1722 (1800 A.D.).

BIBLIOGRAPHY

PRIMARY SOURCES

A. Compiled and Edited Works on Inscriptions

Inscriptions of Ancient Assam (ed.), M.M. Sharma, Guwahati University, 1978.
Kamarupa Sasanawali (ed.), D. Sharma, Guwahati, 1981
Prachya Sasanawali (ed.), M. Neog, Guwahati, 1974.

B. Archaeological and Numismatic

1. A number of remains of the medieval period have been scattered all over the state of Assam in the form of temples, buildings (a few) and the *maidams* or the royal Ahom graveyards at Charaideo in Sibsagar.
2. A number of Koch and Ahom coins have been discovered and preserved by individuals and government institutions like the Assam State Museum, Guwahati; the Indian Museum, Calcutta; the Asiatic Society, Calcutta; the Department of Historical and Antiquarian Studies (DHAS), Guwahati etc.
3. Work on Archaeology: Archaeological Survey of India, *Annual Reports, Archaeology in Assam*, Government of Assam.
4. ***Works on Coins:***
   Catalogue of the Provincial Coin Cabinet, Assam (ed), A.W. Botham, 2nd edn, Allahabad, 1930.
   Catalogue of the Coins in the Indian Museum, Vol. I, by V.A. Smith, Calcutta, 1909.
   Catalogue of the Provincial Cabinet of Coins Eastern Bengal and Assam by H.E. Stappleton, Shillong, 1911.

C. Literary Works

Section I: Unpublished

(a) *Assamese Buranjis or Chronicles:*

Asam Buranji by Padmeswar Singha Buragohain Naobaicha Phukan. It describes the events covering the entire Ahom period. Transcript No. 109, *DHAS*, Guwahati.

Asam Buranji describes the events of the period from Sutupha (1364-1376) to Suhungmung *alias* Dihingia Raja (1497-1539). Tr. No. 2, Vol. V, Pt. I, *DHAS*, Guwahati.

Asam Buranji, a history of Kamarupa with references to its relations with the neighbouring kingdoms. Tr. No. 22, Vol. V, Pt. II, *DHAS*, Guwahati.

Asam Buranji a chronicle describing the events of the period from Suhungmung to Gadadhar Singha (1681-1696). Tr. No. 7, Vol. V, Pt. VII, *DHAS*, Guwahati.

Asam Buranji describes the events from the reign of Suklenmung *alias* Gargaya Raja (1539-1552) to Siva Singha (1714-1744). Kamarupa **Anusandhan Samiti, Guwahati.**

(b) *Charit Puthis or biographies of the Vaishnava Saints:*

Gopal Atar Charit by Ramananda Das, a biography of Gopal Ata of Bhavanipur, a disciple of Madhavadeva.

Guru Charit by Damodar Das, a small biography of Sankaradeva.

Section II: Published

(a) *Buranjis or Chronicles:*

(i) Assamese:

Asam Buranji or *Sri Sri Svarganarayan Maharajar Janma Charitra*, (ed.), S.K. Bhuyan, 2nd edn., Guwahati, 1960.

Asam Buranji by Harakanta Barua Sadaramin, (ed.), S.K. Bhuyan, 2nd edn, Guwahati, 1962.

Asam Buranji Sar by Kashinath Tamuli Phukan. Original title *Asam Buranji Puthi* (Sibsagar, 1844), (ed.), P.C. Choudhury, Guwahati, 1964.

Asam Buranji, obtained from the family of Sukumar Mahanta of North Guwahati, (ed.), S.K. Bhuyan, 3rd edn, Guwahati, 1988.

Asamar Padya Buranji, (a collection of two metrical chronicles by Dutiram Hazarika and Bisweswar Vaidyadhip, (ed.), S.K. Bhuyan, 2nd edn, Guwahati, 1984.

Asam Buranji by Gunabhiram Barua, Reprint, Guwahati, 1972.

Changrung Phukanar Buranji, (ed.), B.K. Barua, Dibrugarh University, 1992.

Deodhai Asam Buranji, (ed.), S.K. Bhuyan, Reprint, Guwahati, 1962.

Jayantiya Buranji, (ed.), S.K. Bhuyan, 2nd edn, Guwahati, 1964.

Kachari Buranji, (ed.), S.K. Bhuyan, 2rd edn, Guwahati, 1984.

Kamrupar Buranji, (ed.), S.K. Bhuyan, 2nd edn, Guwahati, 1958.

Padshah Buranji, (ed.), S.K. Bhuyan, Guwahati, 1935.

Purani Asam Buranji, (ed.), H.C. Goswami, Reprint, Guwahati, 1977.

Satsari Asam Buranji, (a collection of seven old chronicles) (ed.), S.K. Bhuyan, 2nd edn, Guwahati, 1964.

Tripura Buranji (ed.), S.K. Bhuyan, 3rd edn, Guwahati, 1983.

Tungkhungia Buranji by Srinath Duara Barbarua (ed.), S.K. Bhuyan, 2nd edn, Guwahati, 1964.

(ii) Bengali:

Asam Buranji by Haliram Dhekial Phukan, Reprint, Guwahati, 1962.

(iii) English:

Ahom Buranji, (tr. & ed.), Raisahib G.C. Barua, Reprint, Guwahati, 1985.

An Account of Assam, (a translated version of two local chronicles) (tr.), J.P. Wade, (ed.), B. Sharma, North Lakhimpur, 1927.

Annals of the Delhi Badshahate, being a translation of the Assamese chronicle *Padshah Buranji*, (ed. & tr.) S.K. Bhuyan, Guwahati, 1947.

Tungkhungia Buranji, being the translation of the Assamese chronicle *Tungkhungia Buranji*, (tr.), S.K. Bhuyan, 2nd Print, Guwahati, 1968.

Weissalisa, (tr. & ed.), B. Barua and M. Bora respectively, Dibrugarh University, 1977.

(b) *Charit Puthis:*

Aniruddhadevar Charitra Aru Mayamara Satrar Vamsavali, of Chidananda Goswami, Dibrugarh, 1931.

Bardowa Guru Charit, (ed.), M. Neog, Guwahati, 1977.

Guru Charit by Ramcharan Thakur, (ed.), H.N. Dutta Baruah, 3rd edn, Nalbari, 1978.

Guru Charit Katha, (ed.), M. Neog, Guwahati, 1986.

Katha Guru Charit, (ed.), U.C. Lekharn, Nalbari, 15th edn, 1987.

Mahapurusha Sri Sankaradeva Aru Sri Madhavadeva by Daityari Thakur, (ed.), H.N. Dutta Baruah, Nalbari, 1978.

Mahapurusha Sri Sri Gopaldevar Charitra, (ed.), M. Mahanta and N. Bordoloi, Jorhat, 1978.

(c) *Vamsavalis or Genealogical Works:*

Darrang Raj Vamsavali by Suryakhari Daivajna, (ed.), N.C. Sarma, Pathsala, 1973.

(d) *Religious Works:*

(i) Assamese:

Astadasa Parba Mahabharata (2 Vols.) by Ram Saraswati and others, (ed.), H.N. Dutta Barua, Nalbari, 1967.

Kirtan Aru Namghosha by Sankaradeva and Madhavadeva respectively, 12th Reprint, Nalbari, 1975.

Srimad Bhagavata by Sankaradeva and others, (ed.), H.N. Dutta Barua, 8th edn, Nalbari, 1976.

(ii) Sanskrit:

Kalika Puranam, (ed.), Panchanan Tarkaratna, Calcutta, B.S. 1384, (1977-78).

Yogini Tantram, (ed.), Swami S. Saraswati, Calcutta, B.S. 1385, (1979-80).

(e) *Other Works:*

Kumara Harana by Sridhar Kandali, (ed.), L. Gogoi, Sibsagar, 1972.

Rukmini Harana Kavya by Sankaradeva, (ed.), H.N. Dutta Barua, Nalbari, 1947.

(f) *Periodicals and Aphorisms:*

Asam Bandhu, published and edited by Gunabhiram Barua in 1885-1886 (compiled and edited), N. Saikia, Guwahati, 1984.

The Orunodoi, a monthly magazine published by the American Baptist Mission in Assamese from Sibsagar, 1846-1854 (Compiled & edited), M. Neog, Guwahati, 1983.

Sampurna Dakar Bachan (aphorisms of Dakpurusha), (ed.), H. Barman, Guwahati, 1977.

D. Persian Chronicles

Ain-i-Akbari (3 vols.) by Shaikh Abul Fazal Allami, Vol. I, (tr.), H. Blochmann and revised by D.C. Phillot, Calcutta, 1927; Vols. II and III (tr.), H.S. Jarett and revised by Sir J.N. Sarkar, Calcutta, 1949 and 1948.

Alamgirnama by Mirza Muhammad Qazim. A summary translation of the account of Assam and her people as given in the work was done by H. Vansittart in the *Asiatic Researches*, Vol. II, New edition, New Delhi, 1979.

Baharistan-i-Ghaybi by Mirza Nathan *alias* Alla-ud-din Isphani, (tr. & ed.), M.I. Bora, 2 Vols., Guwahati, 1936.

Fathiya-i-Ibriya also called *Tarikh-i-Asam* and *Tarikh-i-Mulk-i-Asham* by Iban Muhammad Wali Ahmed Shihabuddin Talish (tr. in part) by Sir J.N. Sarkar in *JBORS*, 1915, Vol. I, Pt. II; also in part by H. Blochmann in *JASB*, Vol. XLI, Pt. I, 1872.

Padishahnama by Abdul Hamid Lahori (tr. in part) by H. Blochmann in *JASB*, Vol. XLI, Pt. I, 1872.

Tabaqat-i-Nasiri by Minhaj-ud-din-us-Siraj (tr.), Major H. Raverty (2 Vols.), Reprint, New Delhi, 1970.

Riyaz-us-Salatin by Munshi Ghulam Husain Salim, (tr.), Abdus Salaam, Reprint, Delhi, 1975.

E. Foreign Travellers' Accounts

1. The Account of Ralph Fitch, an English traveller, who visited Koch-Behar and western fringe of the Brahmaputra valley in 1585, is incorporated in *Ralph Fitch: England's Pioneer to India, Burma etc.* (ed.), J.H. Ryley, London, 1899; also in *Early Travels in India* by Sir W. Foster, Oxford, 1921.
2. Letters from Stephen Cacella, a Portuguese traveller who visited lower Assam in 1626-27 are incorporated in the *Early Jesuit Travellers in Central Asia* (1603-1721), by C. Wessels, S.J., The Hague, 1924.
3. *A Relation of an Unfortunate Voyage to the Kingdom of Bangala* by Glanius, a Dutch sailor, who accompanied Mir Jumla in his

expedition to Assam in 1662. Reproduced in *Bengal: Past and Present*, Vol. XXIX, 1925.

F. Reports, Monographs etc.

An Account of Assam (collected during the year 1807-1809), Francis (Buchanan) Hamilton, (ed.), S.K. Bhuyan, 2nd edn, Guwahati, 1963.

A Collection of Treaties: Engagement and Sanads Relating to India and Neighbouring Countries, C.U. Aitchison, Reprint, Delhi, 1983.

A Descriptive Account of Assam by W. Robinson, (1841), Reprint, Delhi, 1975.

A Geographical Sketch of Assam (1800), J.P. Wade, (first published in *Asiatic Annual Registrar* in 1805) has been attached to his *An Account of Assam.*

A Sketch of Assam with Some Accounts of the Hill Tribes, John Butler, London, 1847.

An Historical and Descriptive Account of the Kachari Tribes in the North Cachar Hills with Specimens of Tales and Falklore, C.A. Soppit (1885), Reprint, Shillong, 1901.

A Geographical, Statistical and Historical Account of Hindostan and the Adjacent Countries (2 Vols.), Walter Hamilton (London, 1820), Reprint, Delhi, 1971.

A Memoir of David Scott, Major Adam White, (1832) (compiled), S.K. Bhuyan, Guwahati, 1988.

Report on the Province of Assam, A.J.M. Mills (Calcutta, 1854), Reprint, Guwahati, 1984.

Report on the Eastern Frontier of British India, R.B. Pemberton (Calcutta, 1835), Reprint, Guwahati, 1966.

The History, Antiquities, Topography and Statistics of Eastern India (6 Vols.), M. Martin (1838), Reprint, Delhi, 1976.

Topography of Assam, John M'Cosh (Calcutta, 1837), Reprint, Delhi, 1975.

Travels in Assam (original title *Travels and Adventures in the Province of Assam*, 1855), John Butler, 2nd Reprint, Delhi, 1988.

The North-East Frontier of India, A. Mackenzie (1884), 2nd Reprint, Delhi, 1981. Captain Welsh' *Report on Assam*, 1794, with *David Scott's remarks* on it has been incorporated here in Appendix-A.

SECONDARY SOURCES

(i) Books

(a) *Assamese*

Barbarua, H., *Ahomar Din*, Guwahati, 1981.
Barua, B.K., *Asamar Loka Sanskriti*, Guwahati, 1961.
Barua, H.C., *Hemkosh* (ed.), A.R. Barua, 4th edn, Guwahati, 1965.
Bargohain, J.N., *Asamar Arthanaitik Itihas*, Jorhat, 1985.
______, *Asamar Sanskritik Itihas*, Jorhat, 1989.
Bhuyan, N.C., *Baro Bhuyanr Chamu Buranji*, Jorhat, 1961.
Bhuyan, S.K., *Asam Buranjir Katha*, Guwahati, 1989.
_____, *Lachit Barphukan*, Guwahati, 1971.
_____, *Swargadeo Rajeswar Singha*, Guwahati, 1975.
Buragohain, P.K., *Ahmar Adi Buranji*, Jorhat, 1946.
Dutta Barua, H.N., *Prachin Kamrupar Kayastha Samajar Itivritta*, Nalbari, 1941.
Gohain, H., *Kirtan Puthir Rashbichar*, Guwahati University, 1981.
_____, *Asamiya Jatiya Jivanat Mahapurushiya Parampara*, Guwahati, 1987.
Gohain Barooah, P.N., *Asamar Buranji*, Reprint, Guwahati, 1976.
Gogoi, L., *Tai-Sanskritir Ruprekha*, 3rd edn, Calcutta, 1985.
Guha A., *Jamidarkalin Goalpara Jilar Artha—Samajik Avastha*, Dhubri, 1984.
Kakati, B., *Prachin Kamrupar Dharmar Dhara*, Pathsala, 1953.
Mahanta, P., *Asamiya Madhyvitta Srenir Itihas*, Guwahati, 1991.
Narzi, B., *Bodo Kacharir Samaj aru Sanskriti*, Guwahati, 1971.
Neog, M., (comp. & ed.), *Pabitra Asam*, 2nd edn, Jorhat, 1969.
_____, *Purani Asamiya Samaj Aru Sanskriti*, Reprint, Guwahati, 1971.
_____, *Maheswar Neog Rachanawali*, Vol. I, (ed), N. Saikia, Dibrugarh, 1986.
Rajkumar, S., *Itihase Sowara Chasata Bachar*, Jorhat, 1980.
_____, *Chutiya, Bhuyan Aru Matak Rajya*, Guwahati, 1965.
Sharma, B., *Dakshinpat Satra*, Guwahati, 1978.
_____, *Maniram Dewan*, Guwahati, 1950.
_____, *Benudhar Sharma Rachanawali*, 4 Vols. (ed), A.K. Sharma, Guwahati, 1987.
Sharma, G., *Asamiya Jatir Itivritta*, Jorhat, 1974.
Sharma, D., *Mangaldair Buranji*, Guwahati, 1974.
Sarma, T., *Auniati Satrar Buranji*, Auniati Satra, 1975.

(b) *English*

Ali, M. Athar, *The Mughal Nobility Under Aurangzeb*, Reprint, New Delhi, 1970.

Aziz, A., *The Mansabdari System and the Mughal Army*, Lahore, 1945.

Baden-Powell, B.H., *The Land Systems of British India*, (3 Vols.), Vol. III, Indian Reprint, Delhi, 1974.

_____, *Indian Village Community*, India Reprint, Delhi, 1972.

Ball, V., (ed.), Tavernier's *Travels in India*, London, 1925.

Barua, B.K., *A Cultural History of Assam*, Vol. I, 2nd edn., Guwahati, 1969.

Barua, K.L., *Early History of Kamarupa*, 2nd edn, Guwahati, 1966.

Bhuyan, S.K., *Lachit Barphukan and His Times*, Guwahati, 1947.

Barua, K.L., *Studies in the History of Assam* (Comp. & ed.), M. Neog, Jorhat, 1973.

Barua, S.L., *A Comprehensive History of Assam*, Delhi, 1985.

_____, *Last Days of Ahom Monarchy*, Delhi, 1991.

Barpujari, H.K., (ed.), *The Comprehensive History of Assam*, Vols. II, III, Guwahati, 1992, 1994.

_____, *Assam in the Days of the Company, 1826-1858*, Guwahati, 1980.

Banerjee, A.C., *The Eastern Frontier of British India (1784-1826)*, Calcutta, 1943.

_____, *The Agrarian System of Bengal*, (2 Vols.), Calcutta, 1980.

Banerjee, R.D., *History of Orissa*, (2 Vols.), Reprint, Delhi, 1980.

Basu, N.K., *Assam in the Ahom Age*, Calcutta, 1970.

Beal, S., (tr.), *Life of Hiuen Tsiang*, Reprint, New Delhi, 1973.

Beteille, A., *Studies in Agrarian Social Structure*, Reprint, Delhi, 1979.

Bhattacharjee, J.B., *Kachar under the British Rule in North-East India*, Delhi, 1979.

Bhattacharyya, S.N., *A History of Mughal North-East Frontier Policy*, Calcutta, 1929.

Bhuyan, S.K., *Anglo Assamese Relations*, Guwahati, 1948.

_____, *Atan Buragohain and His Times*, Guwahati, 1957.

_____, *Studies in the History of Assam*, Guwahati, 1965.

Brown, R., *Statistical Account of Manipur*, Calcutta, 1873.

Chandra, S., *Medieval India*, New Delhi, 1978.

Chatterjee, S.K., *Kirata-Jana Kriti*, 2nd edn, Calcutta, 1974.

_____, *The Place of Assam in the History and Civilization of India*, 2nd edn, Guwahati University, 1970.

Chopra, P.N., *Some Aspects of Society and Culture During the Mughal Age*, 2nd edn, Agra, 1963.

Choudhury, P.C., *The History of the Civilization of the People of Assam*, 2nd edn, Guwahati, 1966.

Dalton, E.T., *Tribal History of Eastern India* (originally *Descriptive Ethnology of Bengal*, Calcutta, 1872). Reprint, New Delhi, 1973.

Das, J.N., (ed.), *Land System of North-Eastern India*, Vol. VI, (MS).

Dutta, K.K., *Survey of India's Social Life and Economic Conditions in the Eighteenth Century*, Calcutta, 1961.

Devi, L., *Ahom Tribal Relations*, Guwahati, 1968.

Dey, U.N., *The Mughal Government* (1556-1707), Delhi, 1970.

Duncan, W.A.M., *Dye and Dying in Assam*, London, 1895.

Elliot and Dowson, *The History of India as told by it's Own Historians*, Vols. VI & VII, Indian edn, Allahabad, 1964.

Endle, S., *The Kacharis*, Indian Reprint, Delhi, 1975.

Farooque, A.K.M., *Roads and Communications in Mughal India*, Delhi, 1977.

Fitzsimmons, T., (ed.), *Thailand*, New Haven, 1956.

Frykenbery, R.E., (ed.), *Land Control and Social Structure in Indian History*, London, 1969.

Gait, E.A., *A History of Assam*, 2nd edn, Reprint, Guwahati, 1981.

_____, *Report on the Progress of Historical Research in Assam*, Shillong, 1897.

Ghosal, S.C., *A History of Koch Behar* (a translation of Khan Choudhury Amanat-ullah's Bengali work *Koch Beharer Itihas*), Cooch Behar, 1942.

Ghosal, U.N., *The Agrarian System of Ancient India*, Calcutta, 1930.

Gogoi, P., *The Tai and the Tai Kingdoms: With a Fuller Treatment of the Tai Ahom Kingdom in the Brahmaputra Valley*, Guwahati University, 1968.

Gohain, U.N., *Assam Under the Ahoms*, Jorhat, 1942.

_____, *A Short Note on the Tai*, Jorhat, 1950.

Goswami, A., (ed), *Land Reforms and Peasant Movement*, Guwahati, 1986.

Goswami, H.C., *Descriptive Catalogue of Assamese Manuscripts*, Calcutta University, 1930.

_____, *Descriptive Catalogue of Sanskrit Manuscripts*, Guwahati, 1961.

Goswami, S., *Aspects of Revenue Administration in Assam*, Delhi, 1987.

Graham, D.A., *Siam*, Vol. I, London, 1924.

Griffith, W., *Journals of Travels in Assam, Burma, Bhutan and Afghanistan*, 2 Vols., Reprint, Delhi, 1981.

Grierson, G.A., *Linguistic Survey of India*, Vol. 2, Calcutta, 1904.

Guha, A., *Medieval and Early Colonial Assam, Society, Polity, Economy*, Calcutta, 1991.

Gurdon, P.R.T., *The Khasis*, Reprint, Delhi, 1975.

Habib, I., *The Agrarian System of Mughal India, (1556-1707)*, Mumbai, 1963.

Hall, D.G.E., *A History of South East Asia*, 4th edn, Reprint, London, 1987.

Hasan, N., *Thoughts on Agrarian Relations in Mughal India*, New Delhi, 1973.

Hodgson, B.H., *Essay the first on the Koch, Bodo and Dhimal Tribes*, Calcutta, 1847.

Hunter, W.W., *A Statistical Account of Assam*, 2 Vols., Reprint, Delhi, 1990.

Kosambi, D.D., *An Introduction to the Study of Indian History*, Bombay, 1956.

Lahiri, R.M., *Annexation of Assam*, Reprint, Calcutta, 1975.

Latourette, K.S., *The Chinese—Their History and Culture*, 4th revised edn, London etc., 1972.

Maity, S.K., *Economic Life of Northern India in the Gupta Period*, 2nd edn, Delhi, 1970.

Moreland, W.H., *The Agrarian System of Moslem India*, Reprint, Delhi, 1968.

_____, *From Akbar to Aurangzeb*, Reprint, Delhi, 1972.

Mukherjee, R., *The Economic History of India, 1600-1800*, Allahabad, 1967.

Nath, D., *History of the Koch Kingdom (1515-1615)*, Delhi, 1989.

Neog, M., *Sankaradeva and His Times*, Guwahati University, 1965.

_____, *Socio-Political Events in Assam Leading to the Militancy of the Mayamariya Vaishnavas*, Calcutta, 1982.

Ojha, P.N., *Some Aspects of North Indian Social Life, 1556-1707*, Patna, 1961.

Phukan, J.N., *The Economic History of Assam under the Ahoms*, (MS), A Ph.D. thesis, Guwahati University, 1973.

Playfair, Major A., *The Garos*, Reprint, Guwahati, 1975.

Rajguru, S., *Medieval Assamese Society*, Nagaon, 1988.

Ray, A., *Some Aspects of Mughal Administration*, New Delhi, 1984.

Ray, N.R., *Sources of the History of India*, Vol. III, Calcutta, 1988.

Raychoudhuri, H.C., *Political History of Ancient India*, 7th edn, Calcutta, 1972.

Raychoudhuri, T. and Habib, I., *The Cambridge Economic History of India*, 2 Vols., Reprint, Delhi, 1984.

Raychoudhuri, T., *Bengal under Akbar and Jahangir*, Reprint, Delhi, 1969.

Sarkar, Sir J.N., *History of Bengal*, Vol. II, Reprint, Patna, 1973. *Mughal Administration*, Reprint, Calcutta, 1972.

Samman, H.F., *Cotton Fabrics of Assam*, Calcutta, 1897.

Sarkar, J.N., *Mughal Economy: Organisation and Working*, Calcutta, 1987.

_____, *The Life of Mir Jumla: The General of Aurangzeb*, 2nd edn, New Delhi, 1979.

Sharma, S.N., *A Socio-Economic and Cultural History of Medieval Assam (1200-1800)*, Guwahati, 1989.

_____, *The Neo-Vaishnavite Movement and the Satra Institution of Assam*, Guwahati University, 1966.

Sharma, R.S., (ed.), *Land Revenue in India: Historical Studies*, Delhi, 1971.

_____, *Social Changes in Early Medieval India*, Reprint, New Delhi, 1981.

_____, *Indian Feudalism*, 2nd edn, Delhi, 1980.

Shakespeare, L.W., *History of Upper Assam, Upper Burma and North Eastern Frontier*, London, 1914.

Siddiqui, N.A., *Land Revenue Administration under the Mughals'*, Mumbai, 1970.

Sircar, D.C., *Land System and Feudalism in Ancient India*, Calcutta University, 1966.

Sen, S., *Agrarian Relations in India (1793-1947)*, New Delhi, 1979.

Thapar, R., *Ancient Indian Social History*, Reprint, Delhi, 1984.

Vasu, N.N., *The Social History of Kamarupa* (3 Vols. combined), Reprint, Delhi, 1990.

Vinacke, H.M., *A History of Far East in Modern Times*, First Indian edn, New Delhi, 1978.

Wales, H.G.Q., *Ancient Siamese Government and Administration*, London, 1934.

Ward, W.E., *Note on Assam Land Revenue System*, Shillong, 1897.

Watt, G., *Dictionary of Economic Products of India*, Vol. VI, London, 1893.

Wood, W.A.R., *History of Siam*, London, 1926.

(c) *Bengali*

Khan Choudhury Amanatullah, *Koch Beharer Itihas*, Koch Behar, 1926.

Habib, I., *Mughal Bharatar Krishi Vyavastha* (being the Bengali translation of *The Agrarian System of Mughal India (1556-1707)*, tr. Dutta, Pain and Palit, Calcutta, 1985.

Sharma, R.S., *Bharatar Samantatantra* (being the Bengali translation of *Indian Feudalism (300-1200)*, tr. Chattopadhyaya and Dutta, 2nd edn, Calcutta, 1985.

(ii) Gazetteers, Proceedings, Census etc.

Gazetteer of Bengal and North-East India (originally *Provincial Gazetteers of India: Eastern Bengal and Assam*) by B.C. Allen, E.A. Gait, C.G.H. Allen, H.F. Howard, Reprint, Delhi, 1979.

Assam District Gazetteers, B.C. Allen, Shillong, 1905-1907.

Imperial Gazetteers of India, Bengal and Assam, Vols. VI, VIII, IX, Oxford, 1908.

Proceedings of the Indian History Congress.

Proceedings of the North-East India History Association

(iii) Journals

(a) *Assamese*

Amar Pratinidhi

Banhi

(b) *English*

Assam Economic Journal, Dibrugarh University.

Journal of the Asiatic Society of Bengal.

The Journal of the Assam Research Society.

Journal of Historical Research, Dibrugarh University.

Journal of Royal Asiatic Society.

Journal of Bihar and Orissa Research Society.

Indian Economic and Social History Review.

INDEX